Beast ACADEMY

By Art of Problem Solving

MATH GUIDE 4C

JASON BATTERSON
ERICH OWEN

Published by: AoPS Incorporated
10865 Rancho Bernardo Rd Ste 100
San Diego, CA 92127-2102
info@BeastAcademy.com

ISBN: 978-1-934124-54-3

Written by Jason Batterson
Illustrated by Erich Owen
Colored by Greta Selman

Visit the Beast Academy website at www.BeastAcademy.com.
Visit the Art of Problem Solving website at www.artofproblemsolving.com.
Printed in the United States of America.
2017 Printing.

Become a Math Beast!
For additional books,
printables, and more, visit

BeastAcademy.com

This is Guide 4C in a four-book series for fourth grade:

Guide 4A
Chapter 1: Shapes
Chapter 2: Multiplication
Chapter 3: Exponents

Guide 4B
Chapter 4: Counting
Chapter 5: Division
Chapter 6: Logic

Guide 4C
Chapter 7: Factors
Chapter 8: Fractions (+&−)
Chapter 9: Integers

Guide 4D
Chapter 10: Fractions (×&÷)
Chapter 11: Decimals
Chapter 12: Probability

Contents:

Characters 6

How to Use This Book 8

Chapter 7: Factors 12

- Factors 14
- The Sieve 22
- Fizz Buzz 31
- Divisibility 32
- Prime Factorization 40

Chapter 8: Fractions 48

- Review 50
- Adding Fractions 56
- Mixed Numbers 58
- Rummy $^7/_{56}$ 63
- Adding Mixed Numbers 64
- Story Problem 67
- Fraction Subtraction 68
- Mental Math 73

Chapter 9: Integers 78

Negatives 80

Less Than Zero 82

Adding Integers 87

Grogg's Notes 93

Subtracting Integers 94

Rearranging 100

Lizzie's Notes 109

Index 110

Lizzie
"The Bookworm"
Learning to breathe fire
at dragon school
on the weekends
Alex
"The Executive"
Knows seven different ways
to tie a tie!
Grogg (me)
"The Denominator!"
Awesome at: thumb wrestling, cannonballs,
cart racing
Winnie
"The Firecracker"
gets mad at me A LOT!
my mom thinks it's because
she likes me.
I DO NOT!

GrOk
Math lab should probably hire a bodyguard
constantly captured by
calamitous clod
Almost always alliterates
Probably good at juggling
Ms. Q.
Math Teacher
May someday turn into a really big butterfly
Kraken
pirate shop Teacher
sergeant ROte
gym teacher
VR-960A HOversphere
is the same size as a regulation basketball
ROsencrantz and Guildenstern
custOdian(s?)
Have the same birthday!
when One eats, does the Other feel full?
FiOna
math team cOach
fastest 2-legged sprinter in her grade at B.A.

The Headmaster
How to use this book

If you've gotten this far, you probably know a little bit about how to read a comic book.
You read a comic book the same way you read any other book... from left to right and from top to bottom.
But sometimes, panels don't have frames...
...like the open panel I'm in now.
On each page, start in the top left panel.
Go to the right, then down.
Read all of the balloons in each panel from left to right and top to bottom before moving to the next panel.
At first, you may need to think about which balloon to read next.
Like when lots of characters are talking.
Or when a character speaks more than once.
Right! And sometimes several balloons get connected.
With a little practice, reading comics becomes natural.
How many panels are on this page?

Did you notice this stop sign at the bottom of the previous page?
How many panels are on this page?
Always try to answer the questions in the stop signs before reading on.
There were 5 panels on page 9.
What's this gray box for?
STOP SIGNS ASK QUESTIONS THAT YOU SHOULD TRY TO ANSWER BEFORE READING ANY FURTHER.
The gray boxes in the Guide provide important information and definitions.*
Sometimes the gray boxes offer problem solving tips or other useful information.
*AN ASTERISK MAY BE USED TO TIE THE TEXT IN A COMIC BALLOON TO A GRAY BOX.

Some of the pages in this book are notes pages, created by our students.
RECESS
Game Play:
Other pages have games you can play.

Practice: Pages 6, 38, and 72

Contents: Chapter 7

See page 6 in the Practice book for a recommended reading/practice sequence for Chapter 7.

Factors 14
Can you name every factor of 48?

The Sieve 22
How many primes are less than 100?

Fizz Buzz 31
What comes after 89 in a game of Fizz Buzz?

Divisibility 32
Is 5,500,050 divisible by 9?

Prime Factorization 40
What is the prime factorization of 143?

Chapter 7:
Factors

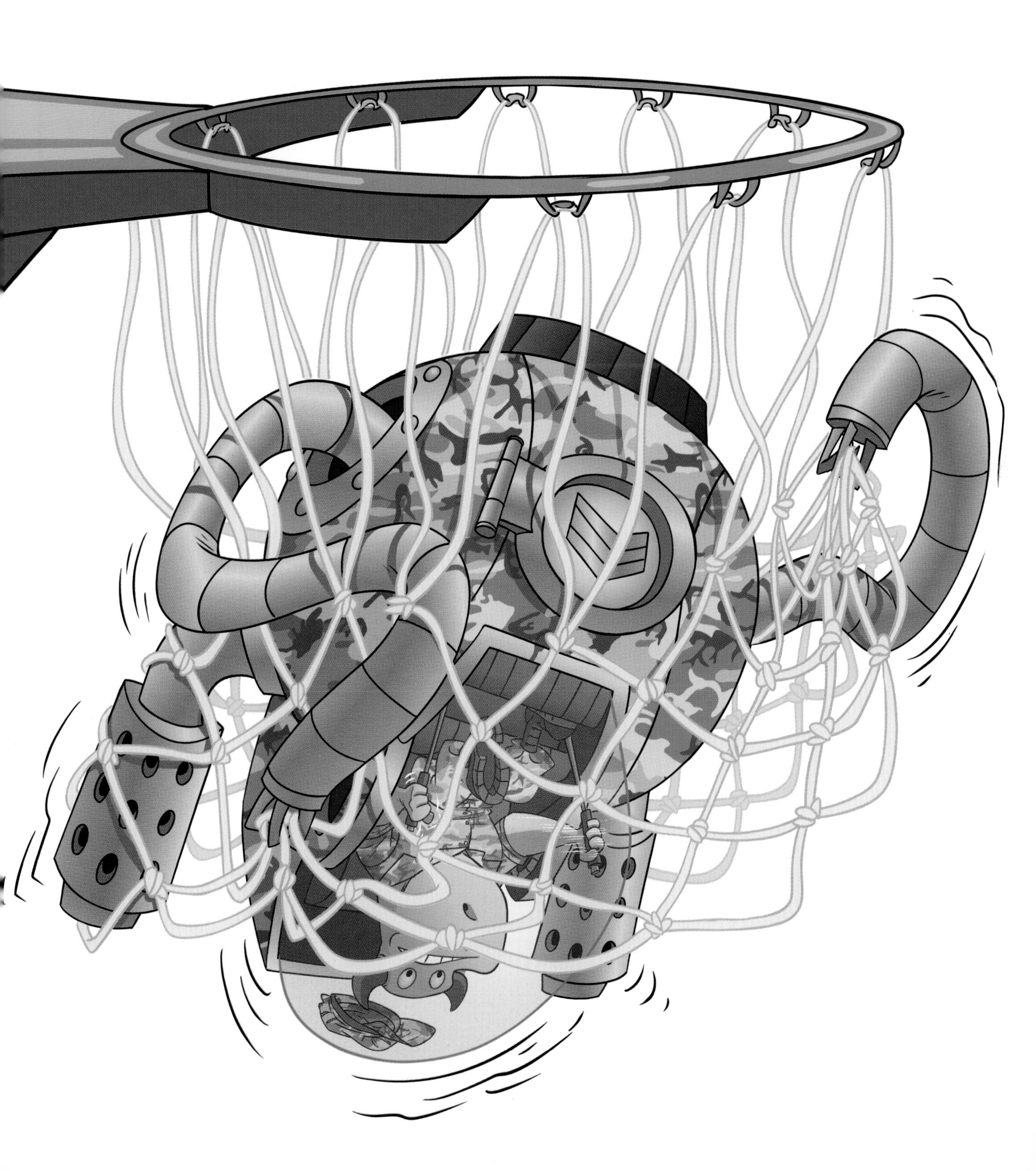

Ms. Q.
Factors
What are factors?
Factors are the numbers that you multiply to get other numbers.
Huh?
For example, when you multiply 2×3, you get 6.
So, we say that 2 and 3 are *factors* of 6.
Factors are the numbers that a number is divisible by.
Since 6 is *divisible* by 3, we say that 3 is a *factor* of 6.
Very good. Does 6 have any other factors?
Hmmm... What other numbers can we multiply to get 6?

I know! 1×6=6. So, 1 and 6 are factors of 6.
That's right.
Six has four factors.
Factors of 6:
1, 2, 3, 6
Let's try a larger number. What are all the factors of 48?
6×8=48, so 6 and 8 are both factors of 48.
1×48=48, so 1 and 48 are both factors.
4×12=48, too. That gives us six factors of 48 so far: 6, 8, 1, 48, 4, and 12.
Did we miss any?
It's hard to tell. We should organize our work.
We can check every number from 1 to 48 to see if it's a factor.
Starting with 1, we know 1×48=48. So, we write 1×48 at the top of our list.
Try 2 next.
48:
1×48
Is 2 a factor of 48? Can you find all of the factors of 48?

2×24=48. We can add 2×24 to our list.
Next, we check 3.
48:
1×48
2×24

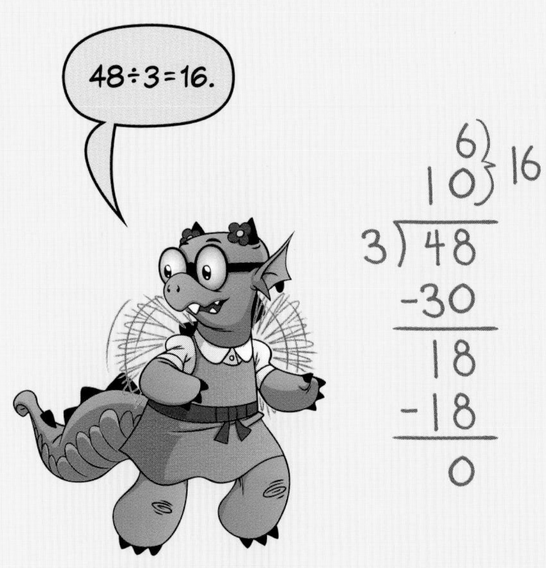
48÷3=16.
6
10
16
3)48
-30
18
-18
0

So, 3×16=48.
We already found 4×12=48, so we can add that to the list.
48:
1×48
2×24
3×16
4×12

48 doesn't end in a 0 or a 5, so 48 isn't divisible by 5.

6×8=48 is next.
48:
1×48
2×24
3×16
4×12
6×8

Then, we check to see if 7 is a factor.

7×6=42, and 7×7=49. So, 48 is not divisible by 7.

48:
1×48
2×24
3×16
4×12
6×8

	quotient	remainder
71÷1	71	0
71÷2	35	1
71÷3	23	2
71÷4	17	3
71÷5	14	1
71÷6	11	5
71÷7	10	1
71÷8	8	7
71÷9	7	8
71÷10	7	1

NUMBERS THAT ARE PRIME ARE OFTEN SIMPLY CALLED "PRIMES."

POSITIVE MEANS GREATER THAN 0. WE WILL LEARN ABOUT NUMBERS THAT ARE LESS THAN 0 IN CHAPTER 9.

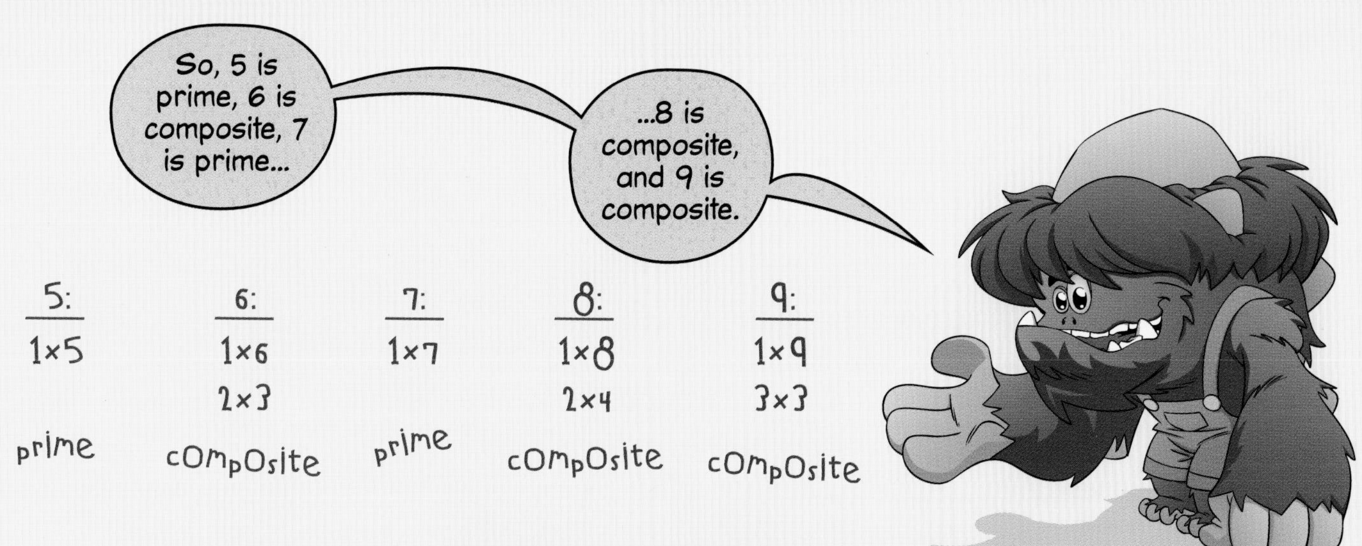
So, 5 is prime, 6 is composite, 7 is prime...
...8 is composite, and 9 is composite.
5: 1×5 prime
6: 1×6 2×3 composite
7: 1×7 prime
8: 1×8 2×4 composite
9: 1×9 3×3 composite

The 1-digit primes are 2, 3, 5, and 7.
Circle the primes:
0, 1, 2, 3, 4, 5, 6, 7, 8, 9

Is 41 prime, too, Ms. Q.?
It is, Grogg. Why do you ask?

HAPPY BIRTHDAY
Last week, my dad turned 41.
I asked him if he felt old. He said...
41
Not me. I'm in my *prime*.

PRINT YOUR OWN 100 CHART AT BEASTACADEMY.COM AND FOLLOW ALONG WITH THE LITTLE MONSTERS.

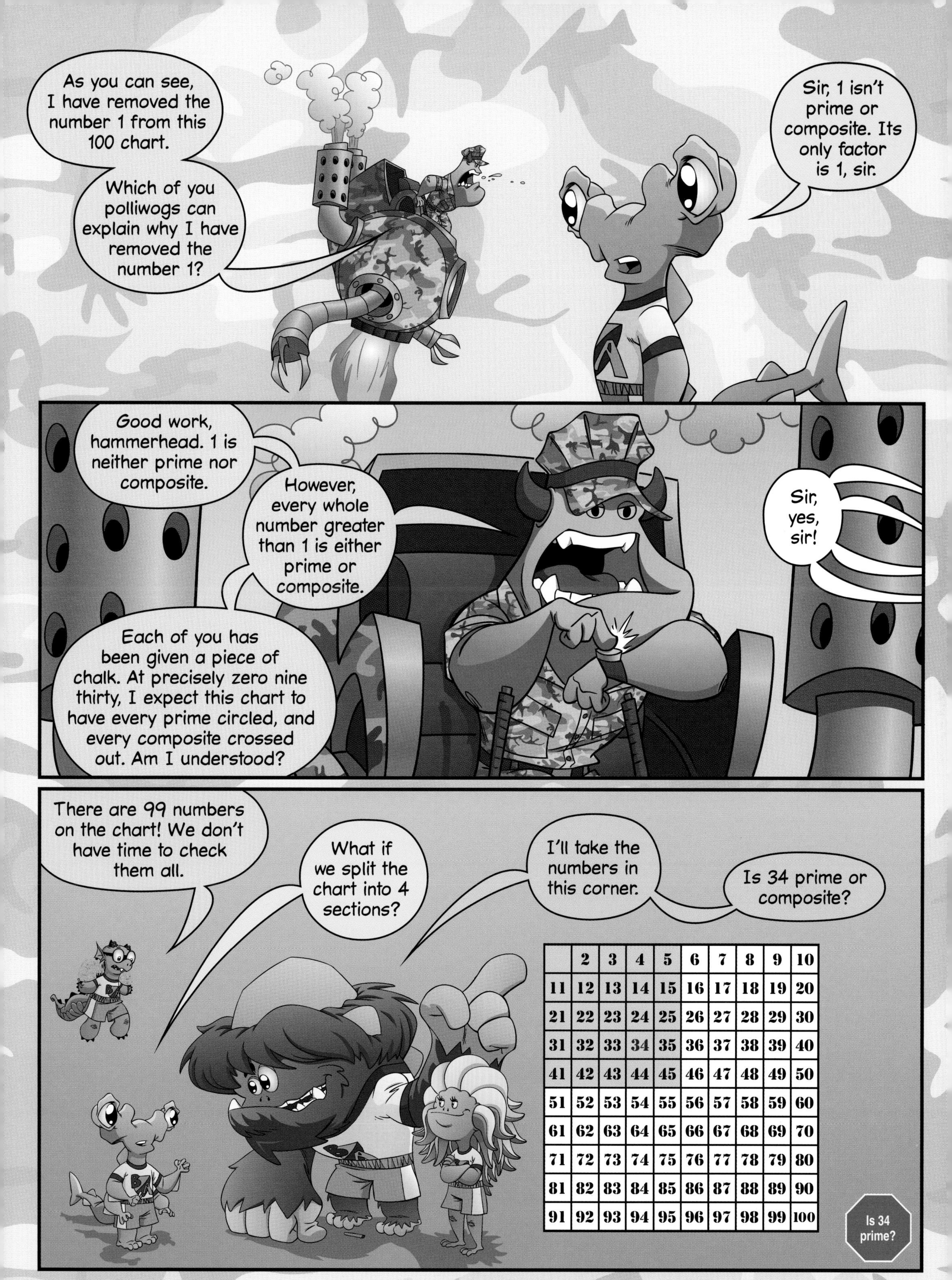

	2	3	4	5	6	7	8	9	10
11	12	13	14	15	16	17	18	19	20
21	22	23	24	25	26	27	28	29	30
31	32	33	34	35	36	37	38	39	40
41	42	43	44	45	46	47	48	49	50
51	52	53	54	55	56	57	58	59	60
61	62	63	64	65	66	67	68	69	70
71	72	73	74	75	76	77	78	79	80
81	82	83	84	85	86	87	88	89	90
91	92	93	94	95	96	97	98	99	100

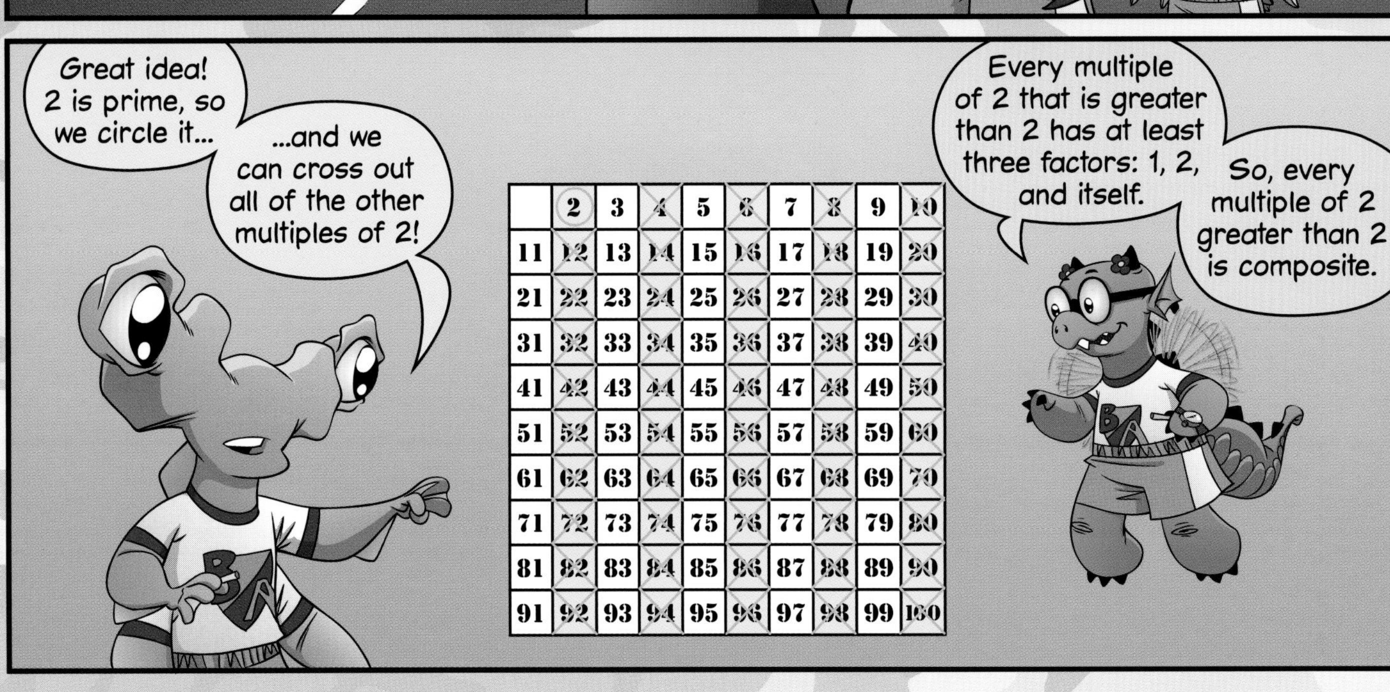

	2	3	4	5	6	7	8	9	10
11	12	13	14	15	16	17	18	19	20
21	22	23	24	25	26	27	28	29	30
31	32	33	34	35	36	37	38	39	40
41	42	43	44	45	46	47	48	49	50
51	52	53	54	55	56	57	58	59	60
61	62	63	64	65	66	67	68	69	70
71	72	73	74	75	76	77	78	79	80
81	82	83	84	85	86	87	88	89	90
91	92	93	94	95	96	97	98	99	100

Sounds good. Let's cross out the multiples of 3.
Except for 3, of course.
Yep. 3 is prime, so we circle it.
And we cross out all the other multiples of 3, which are composite!

Some of the multiples of 3 were already crossed out, like 30.
The multiples of 3 that were already crossed out are also multiples of 2.

Okay... We crossed out multiples of 2 and 3. Next is 4.
It's already crossed out, so we know it's composite.
Do we need to cross out the other multiples of 4?
What do you notice about the multiples of 4?

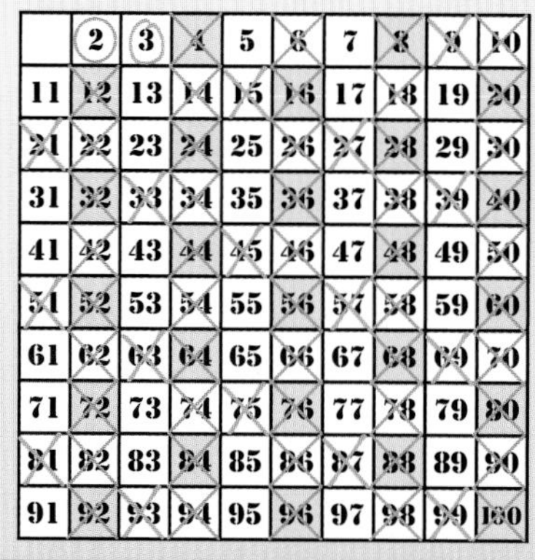

	2	3	4	5	6	7	8	9	10
11	12	13	14	15	16	17	18	19	20
21	22	23	24	25	26	27	28	29	30
31	32	33	34	35	36	37	38	39	40
41	42	43	44	45	46	47	48	49	50
51	52	53	54	55	56	57	58	59	60
61	62	63	64	65	66	67	68	69	70
71	72	73	74	75	76	77	78	79	80
81	82	83	84	85	86	87	88	89	90
91	92	93	94	95	96	97	98	99	100

Five is prime, so we circle 5...

	2	3	4	5	6	7	8	9	10
11	12	13	14	15	16	17	18	19	20
21	22	23	24	25	26	27	28	29	30
31	32	33	34	35	36	37	38	39	40
41	42	43	44	45	46	47	48	49	50
51	52	53	54	55	56	57	58	59	60
61	62	63	64	65	66	67	68	69	70
71	72	73	74	75	76	77	78	79	80
81	82	83	84	85	86	87	88	89	90
91	92	93	94	95	96	97	98	99	100

...and cross out all of the other multiples of 5.

Next is 6.

	2	3	4	5	6	7	8	9	10
11	12	13	14	15	16	17	18	19	20
21	22	23	24	25	26	27	28	29	30
31	32	33	34	35	36	37	38	39	40
41	42	43	44	45	46	47	48	49	50
51	52	53	54	55	56	57	58	59	60
61	62	63	64	65	66	67	68	69	70
71	72	73	74	75	76	77	78	79	80
81	82	83	84	85	86	87	88	89	90
91	92	93	94	95	96	97	98	99	100

6 is already crossed out, so 6 is composite...

...and since 6=2×3, all of the other multiples of 6 got crossed out when we crossed out multiples of 2 and 3.

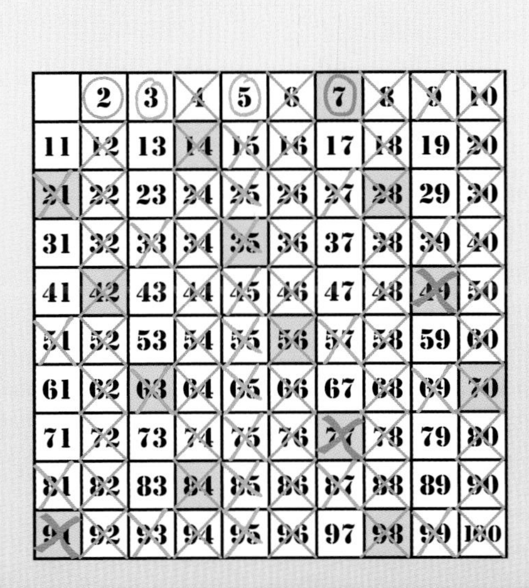

	2	3	4	5	6	7	8	9	10
11	12	13	14	15	16	17	18	19	20
21	22	23	24	25	26	27	28	29	30
31	32	33	34	35	36	37	38	39	40
41	42	43	44	45	46	47	48	49	50
51	52	53	54	55	56	57	58	59	60
61	62	63	64	65	66	67	68	69	70
71	72	73	74	75	76	77	78	79	80
81	82	83	84	85	86	87	88	89	90
91	92	93	94	95	96	97	98	99	100

	2	3	4	5	6	7	8	9	10
11	12	13	14	15	16	17	18	19	20
21	22	23	24	25	26	27	28	29	30
31	32	33	34	35	36	37	38	39	40
41	42	43	44	45	46	47	48	49	50
51	52	53	54	55	56	57	58	59	60
61	62	63	64	65	66	67	68	69	70
71	72	73	74	75	76	77	78	79	80
81	82	83	84	85	86	87	88	89	90
91	92	93	94	95	96	97	98	99	100

	2	3	4	5	6	7	8	9	10
11	12	13	14	15	16	17	18	19	20
21	22	23	24	25	26	27	28	29	30
31	32	33	34	35	36	37	38	39	40
41	42	43	44	45	46	47	48	49	50
51	52	53	54	55	56	57	58	59	60
61	62	63	64	65	66	67	68	69	70
71	72	73	74	75	76	77	78	79	80
81	82	83	84	85	86	87	88	89	90
91	92	93	94	95	96	97	98	99	100

Since we already crossed out multiples of 2, 3, 4, 5, 6, 7, 8, 9, and 10...

...we already crossed out 11×2, 11×3, 11×4, 11×5, 11×6, 11×7, 11×8, and 11×9.

11×10 isn't even on the chart.

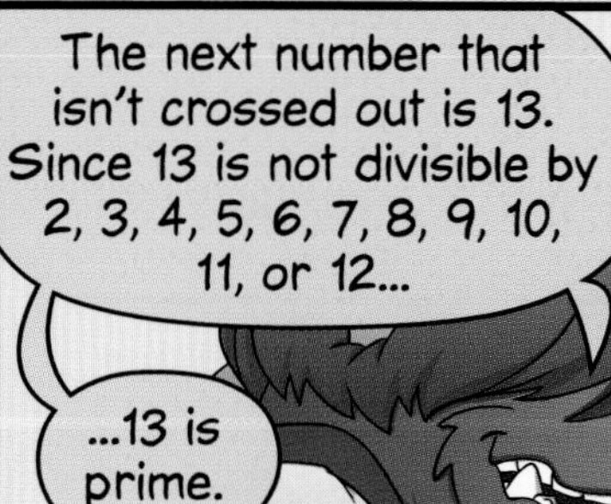

	2	3	4	5	6	7	8	9	10
11	12	13	14	15	16	17	18	19	20
[illegible]	[illegible]	23	24	25	26	27	28	29	30
[illegible]	32	33	34	35	36	37	38	39	40
[illegible]	42	43	44	45	46	47	48	49	50
[illegible]	[illegible]	53	54	55	56	57	58	59	60
61	62	63	64	65	66	67	68	69	70
71	72	73	74	75	76	77	78	79	80
81	82	83	84	85	86	87	88	89	90
91	92	93	94	95	96	97	98	99	100

There aren't any more composite numbers on our chart to cross out!
Huh?
Think about it, Grogg...
...we only need to cross out multiples of primes.

We already crossed out the multiples of 2, 3, 5, and 7.
And the multiples of the larger primes on our chart are already crossed out.
I see.
The next number that isn't crossed out is 17, so 17 is prime. And 17×2, 17×3, 17×4, and 17×5 are already crossed out.

And since we've crossed out every composite number, all of the remaining numbers on our chart are prime.
Let's hurry up and circle them. Our time is almost up.
Fall in, tadpoles!

*THE PROCESS IS CALLED THE SIEVE OF ERATOSTHENES (PRONOUNCED SIV OF AIR-UH-*TOSS*-THE-KNEES).

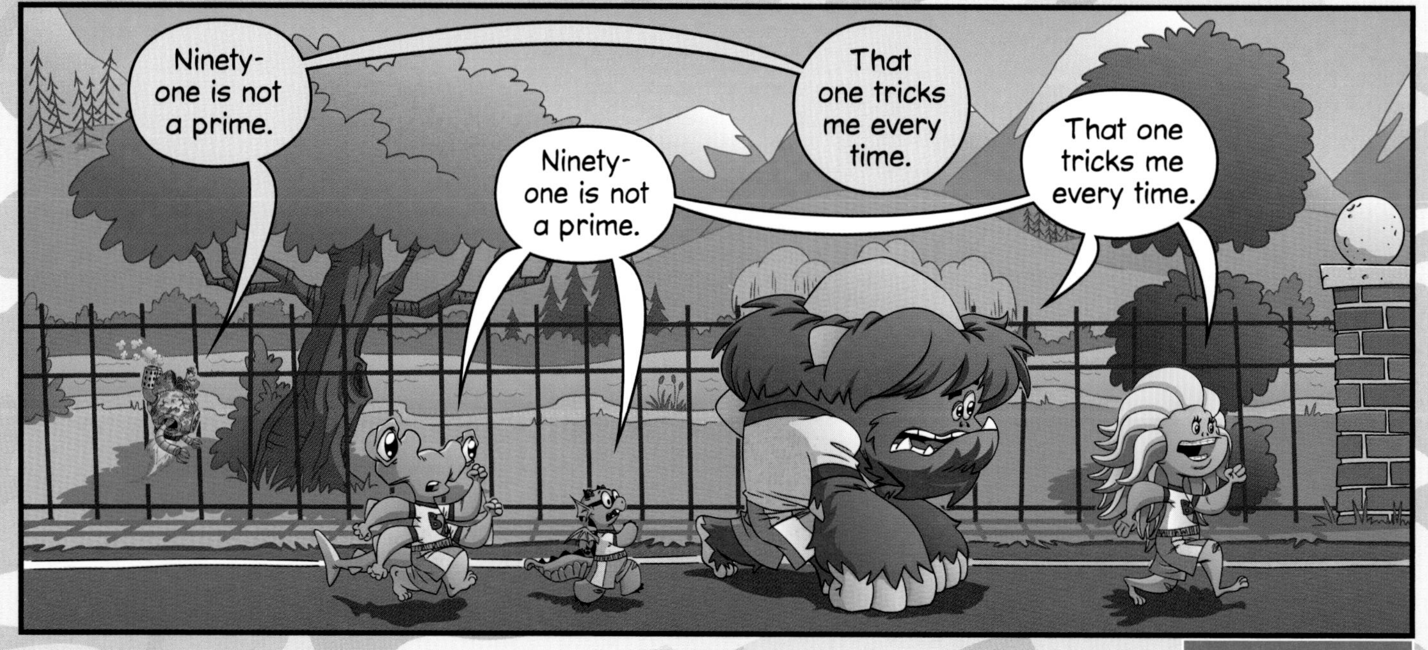

Fizz Buzz

Game Play:

Fizz Buzz is a counting game for two or more players, but is best when played with 5 or more. Players arrange themselves in a circle. Player 1 begins the game by saying "1." The player to the left says "2." Play continues clockwise around the circle with each player stating the next number. However, any multiple of 3 is replaced with the word "fizz," and any multiple of 5 is replaced with the word "buzz." A multiple of both 3 and 5 is replaced with the word "fizzbuzz."

For example, a standard round of Fizz Buzz begins as shown:

1, 2, fizz, 4, buzz, fizz, 7, 8, fizz, buzz, 11, fizz, 13, 14, fizzbuzz, 16, 17, fizz, 19, buzz, fizz, 22, 23, fizz, buzz, ...

Any player who hesitates or makes a mistake is eliminated, and the player to his or her left begins the next round at 1. Play continues until only one player remains. That player is declared the winner.

Variations:

- Different factors can replace 3 and 5. Similarly, different words or gestures can replace the words "fizz" and "buzz." For example, multiples of 3 can be replaced with a clap, and multiples of 7 can be replaced with a hop, with multiples of both 3 and 7 replaced with a clap and a hop.
- Play switches direction on the word fizz or buzz, but not the word fizzbuzz.
- Additional factors with associated words and gestures can be added to make the game even more difficult.
- Instead of restarting at 1 after a mistake, play can resume at the number on which the mistake was made.

Even better when played in a large group!

Cheat Sheet:

Give this list to an adult so they can referee a game of Fizz Buzz between you and your friends.

1	34	67
2	buzz	68
fizz	fizz	fizz
4	37	buzz
buzz	38	71
fizz	fizz	fizz
7	buzz	73
8	41	74
fizz	fizz	fizzbuzz
buzz	43	76
11	44	77
fizz	fizzbuzz	fizz
13	46	79
14	47	buzz
fizzbuzz	fizz	fizz
16	49	82
17	buzz	83
fizz	fizz	fizz
19	52	buzz
buzz	53	86
fizz	fizz	fizz
22	buzz	88
23	56	89
fizz	fizz	fizzbuzz
buzz	58	91
26	59	92
fizz	fizzbuzz	fizz
28	61	94
29	62	buzz
fizzbuzz	fizz	fizz
31	64	97
32	buzz	98
fizz	fizz	fizz

How does this help us see that 72,945 is divisible by 9?

WHEN WE ADD TWO OR MORE MULTIPLES OF A NUMBER, WE GET ANOTHER MULTIPLE OF THE SAME NUMBER. IN THE EXAMPLE ABOVE, ADDING THREE MULTIPLES OF 9 GIVES US ANOTHER MULTIPLE OF 9.

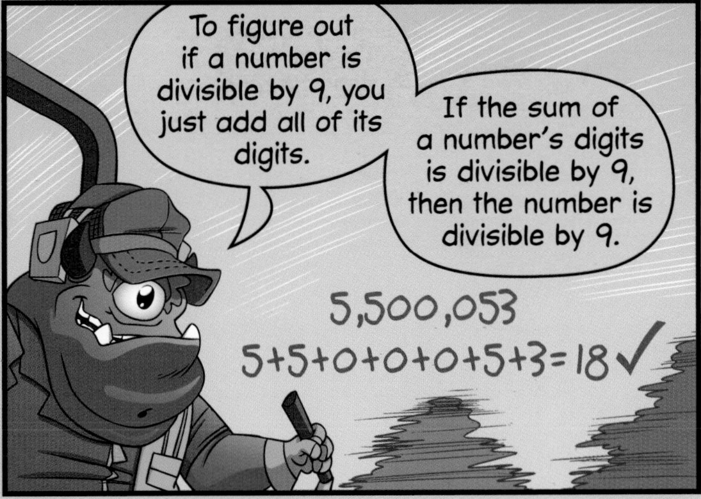

Why does *that* work?
That's the coolest part!
It's kind of tricky to explain.
First, you need to know a little bit about remainders.

What is the remainder when 10 is divided by 9?
1.
Right.
What is the remainder when 20 is divided by 9?
2.
How about 30?
3.

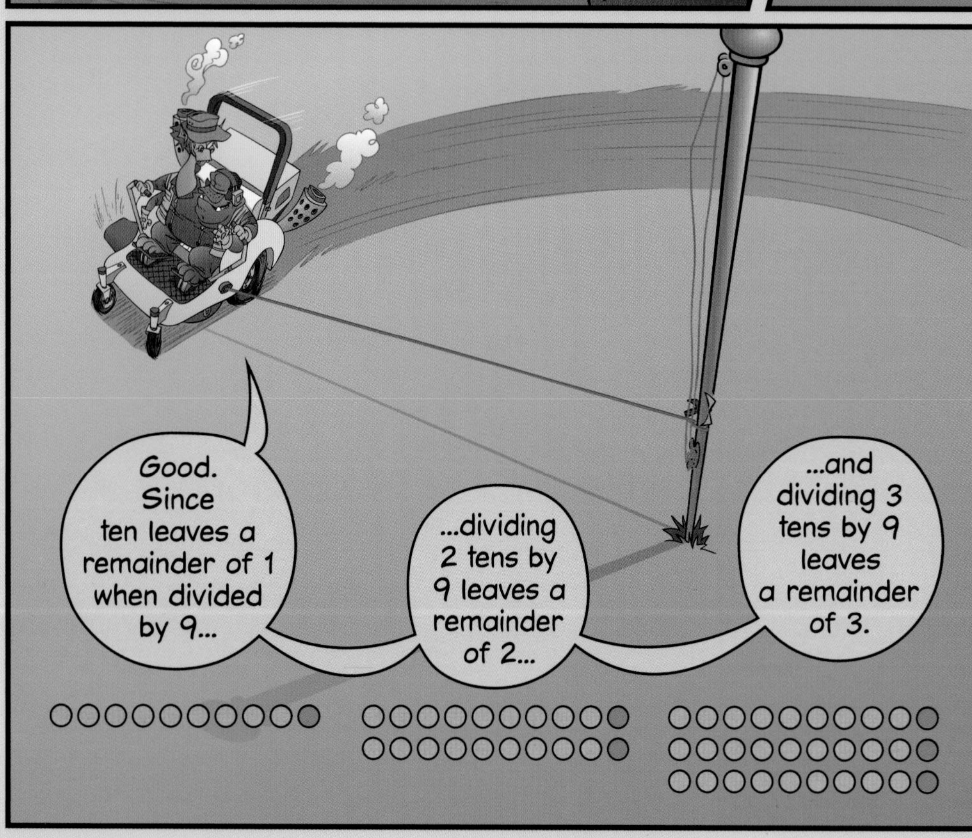
Good. Since ten leaves a remainder of 1 when divided by 9...
...dividing 2 tens by 9 leaves a remainder of 2...
...and dividing 3 tens by 9 leaves a remainder of 3.

What is the remainder when 70 is divided by 9?

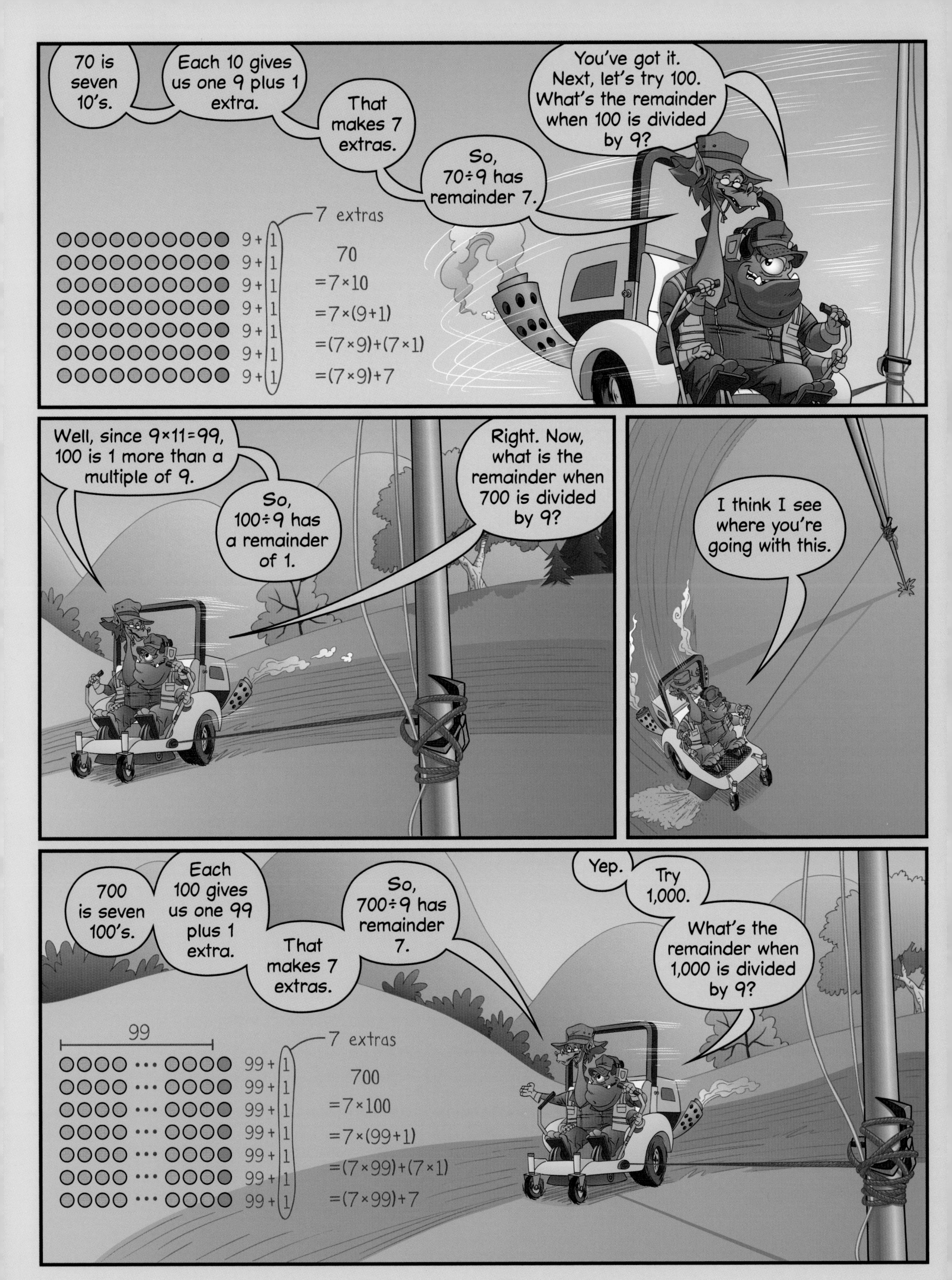
70 is seven 10's.
Each 10 gives us one 9 plus 1 extra.
That makes 7 extras.
So, 70÷9 has remainder 7.
You've got it. Next, let's try 100. What's the remainder when 100 is divided by 9?
7 extras
9 + 1
9 + 1
9 + 1
9 + 1
9 + 1
9 + 1
9 + 1
70
=7×10
=7×(9+1)
=(7×9)+(7×1)
=(7×9)+7
Well, since 9×11=99, 100 is 1 more than a multiple of 9.
So, 100÷9 has a remainder of 1.
Right. Now, what is the remainder when 700 is divided by 9?
I think I see where you're going with this.
700 is seven 100's.
Each 100 gives us one 99 plus 1 extra.
That makes 7 extras.
So, 700÷9 has remainder 7.
Yep.
Try 1,000.
What's the remainder when 1,000 is divided by 9?
99
7 extras
99 + 1
99 + 1
99 + 1
99 + 1
99 + 1
99 + 1
99 + 1
700
=7×100
=7×(99+1)
=(7×99)+(7×1)
=(7×99)+7

Well, since 9×111=999, we know that 1,000 is 1 more than a multiple of 9.
So, 1,000÷9 has a remainder of 1.
Right.
In fact, every power of 10 is 1 more than a multiple of 9!
10,000 is 1 more than 9,999.
100,000 is 1 more than 99,999.
And 1,000,000 is 1 more than 999,999.
Since 10 and 100 and 1,000,000 are each 1 more than a multiple of 9...
...70 and 700 and 7,000,000 are each 7 more than a multiple of 9!
So, 70 and 700 and 7,000,000 each leave a remainder of 7 when divided by 9.
70
=7×10
=7×(9+1)
=(7×9)+(7×1)
=(7×9)+7
700
=7×100
=7×(99+1)
=(7×99)+(7×1)
=(7×99)+7
7,000,000
=7×1,000,000
=7×(999,999+1)
=(7×999,999)+(7×1)
=(7×999,999)+7

It's easy to find the remainder when a number like 400 or 6,000 or 30,000 is divided by 9.
The remainder is the same as the first digit of the number.
400 = 4×100 = 4×(99+1) = 4×99 + 4
6,000 = 6×1,000 = 6×(999+1) = 6×999 + 6
30,000 =3×10,000 =3×(9,999+1) =3×9,999 + 3

Right. Now, try 504.
What's the remainder when 504 is divided by 9?

Hmmm.... I know that 500 is 5 more than a multiple of 9...
...so 504 is 5+4=9 more than a multiple of 9.
500 =5 × 100 =5 × (99+1) =(5×99)+(5×1) =(5×99) + 5
504 =(5 × 99 + 5) + 4 =(5 × 99) + 9

But, if we add 9 to a multiple of 9...
...we get the next multiple of 9!
So, 504 is divisible by 9!
Bingo! How about 765?
Is 765 divisible by 9?

Hmmm....
I know that 700 is 7 more than a multiple of 9.
And 60 is 6 more than a multiple of 9.
So, 765 is 7+6+5=18 more than a multiple of 9.
700
=(7×99)+7
60
=(6×9)+6
765
= 700 + 60 + 5
= (7×99)+7 + (6×9)+6 + 5
= (7×99) + (6×9) + 7+6+5

And since 18 is 9×2, we can write 765 as the sum of three multiples of 9.
So, 765 is divisible by 9!
765
= 700 + 60 + 5
= (7×99)+7 + (6×9)+6 + 5
= (7×99) + (6×9) + 7+6+5
= (7×99) + (6×9) + (2×9)

Adding the digits of a number tells you how much larger the number is than a multiple of 9.

So, if adding the digits gives you a multiple of 9, then the number is a multiple of 9!
For example, 3,888 is 3+8+8+8=27 more than a multiple of 9.
Exactly!
And since 27 is a multiple of 9, 3,888 is also a multiple of 9.
3,888
3+8+8+8=27 ✓

I think we can use a similar trick to see if a number is divisible by 3!

Since 9 and 99 and 999 are all multiples of 9, they're also multiples of 3.
For example, 425 is 4+2+5=11 more than a multiple of 3.
425
= 400 + 20 + 5
= (4×99)+4 + (2×9)+2 + 5
= (4×99) + (2×9) + 4+2+5
So, when we add the digits of a number, we find out how much larger the number is than a multiple of 3.

And since 11 is not a multiple of 3, neither is 425!

That makes sense! I'm gonna write that down.
If the sum of a number's digits is divisible by 9, then the number is divisible by 9.
If the sum of a number's digits is divisible by 3, then the number is divisible by 3.

We're good!

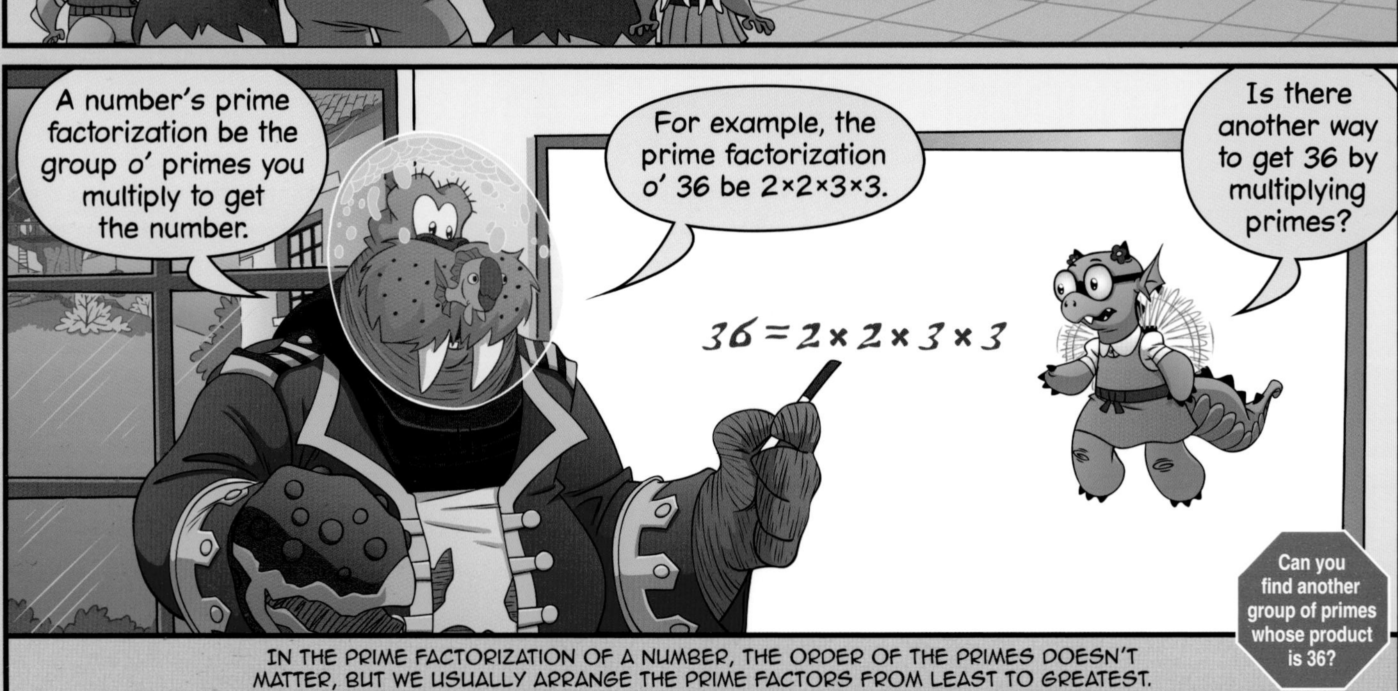

IN THE PRIME FACTORIZATION OF A NUMBER, THE ORDER OF THE PRIMES DOESN'T MATTER, BUT WE USUALLY ARRANGE THE PRIME FACTORS FROM LEAST TO GREATEST.

*0 AND 1 ARE SPECIAL. EVERY PRIME IS A FACTOR OF 0. NO PRIME IS A FACTOR OF 1.

How do we prime factorize a number?
To find the prime factorization of a number, we'll be usin' a factor tree.
"FACTORIZE" IS NOT A REAL WORD.

Does that mean we get to go outside?

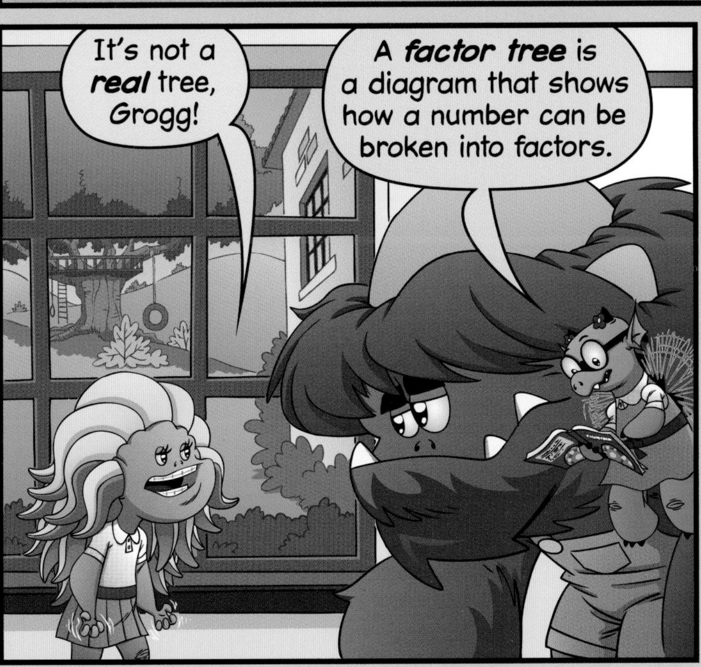
It's not a *real* tree, Grogg!
A ***factor tree*** is a diagram that shows how a number can be broken into factors.

Aye. Let's use a factor tree to find the prime factorization o' 56.
56

First, we look for two numbers that have a product of 56.
Like 7×8?
Exactly.

We write the two factors below the 56, like this.
56
7
8
Can you guess what comes next?

7 is prime, so we've already found one prime factor of 56.
Aye. We'll be circlin' the primes we find in our factor tree.
56
7
8
8 isn't prime. We can factor 8 into 4×2.
56
7
8
4
2
And since 2 is prime, we circle it!
Now you're gettin' the hang of it, lad!
56
7
8
4
2
Then, we can factor 4 into 2×2.
56
7
8
4
2
2
2
And we circle these two 2's because 2 is prime!
The circled numbers give us the prime factors of 56...
...three 2's and a 7.
Aye. We usually write the factors in order, like so.
56
7
8
4
2
2
2
56 = 2×2×2×7

Aren't there other ways to make a factor tree for 56? What if we start with 2×28?
56
2 28

There be many ways to make the tree, but the circled primes always be the same.
56
2 28
7 4
2 2
56=2×2×2×7
SOMETIMES, MATH BEASTS USE EXPONENTS TO WRITE PRIME FACTORIZATIONS. FOR EXAMPLE, 56=2×2×2×7=2³×7.

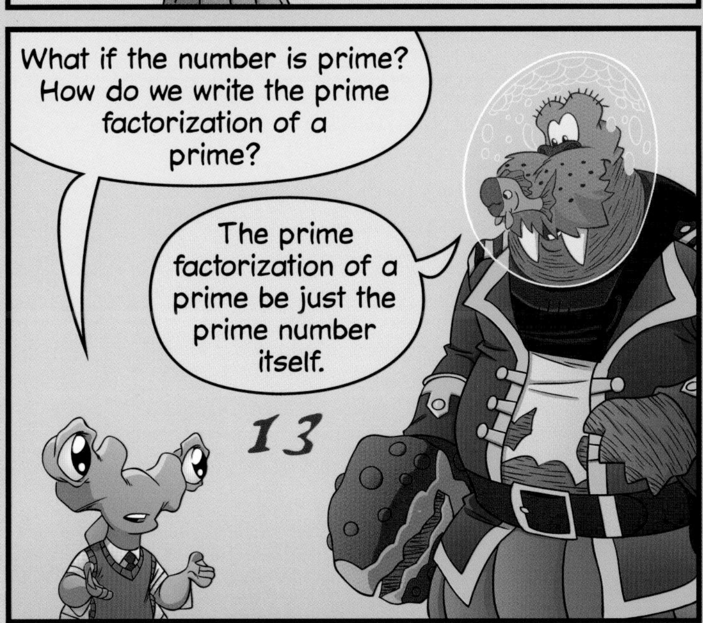
What if the number is prime? How do we write the prime factorization of a prime?
The prime factorization of a prime be just the prime number itself.
13

How do we *know* a number is prime?
Suppose we wanted to write the prime factorization of 123?
How can we tell if 123 is prime?
123

You'll be needin' to see if 123 has any factors, other than 1 and 123.
Start checkin' with 2, 'n' work your way up.
Since 123 is odd, we know that 2 is not a factor of 123.
What is the prime factorization of 123?

Since 1+2+3=6 is divisible by 3, 123 is divisible by 3.
3×40=120. So, 3×41=123!
123
3
41
And both 3 and 41 are prime, so our factor tree is done!
The prime factorization of 123 is 3×41.
123=3×41
WE LEARNED THAT 41 IS PRIME ON PAGE 29.

Aye. Good job. Try another. What be the prime factorization of 143?
143

143 is obviously not divisible by 2.
And since 1+4+3=8 is not divisible by 3, 143 isn't divisible by 3.

What about 4?
Grogg! Since 2 isn't a factor of 143, we already know 4 isn't a factor of 143.
Since 4 is 2×2, every number that has 4 as a factor also has 2 as a factor.
We only need to check for ***prime*** factors.

5 is the next prime.
Since 143 doesn't end in a 0 or a 5, we know 143 is not divisible by 5.
We don't need to check 6!
6=2×3.
Since 2 and 3 are not factors of 143, we know 6 isn't a factor of 143.
Next is 7, and since 7×20=140, we know 143 is 3 more than a multiple of 7.
We don't need to check 8, 9, or 10, because those numbers are all composite.
11 is the next prime.
11×11 is 121.
11×12 is 132.
11×13 is 143!
11×11 = 121
11×12 = 132
11×13 = 143
And since 11 and 13 are both prime, the prime factorization of 143 is 11×13!
143
11
13
Well blow me down!
Try one more.
What be the prime factorization of 163?
What is the prime factorization of 163?

Well, 163 isn't divisible by 2.
Or 3.
Or 5.
How about 7?
163
1+6+3=10
7×20=140. To compute 7×23, we add three more 7's.
7×23=161. So, 163 is 2 more than a multiple of 7.
11 is the next prime. Is 163 divisible by 11?
110 is divisible by 11, so we need 53 more to get to 163.
110+44=154 ✓
110+53=163 ×
110+55=165 ✓
But, since 53 isn't divisible by 11, 110+53=163 isn't divisible by 11.

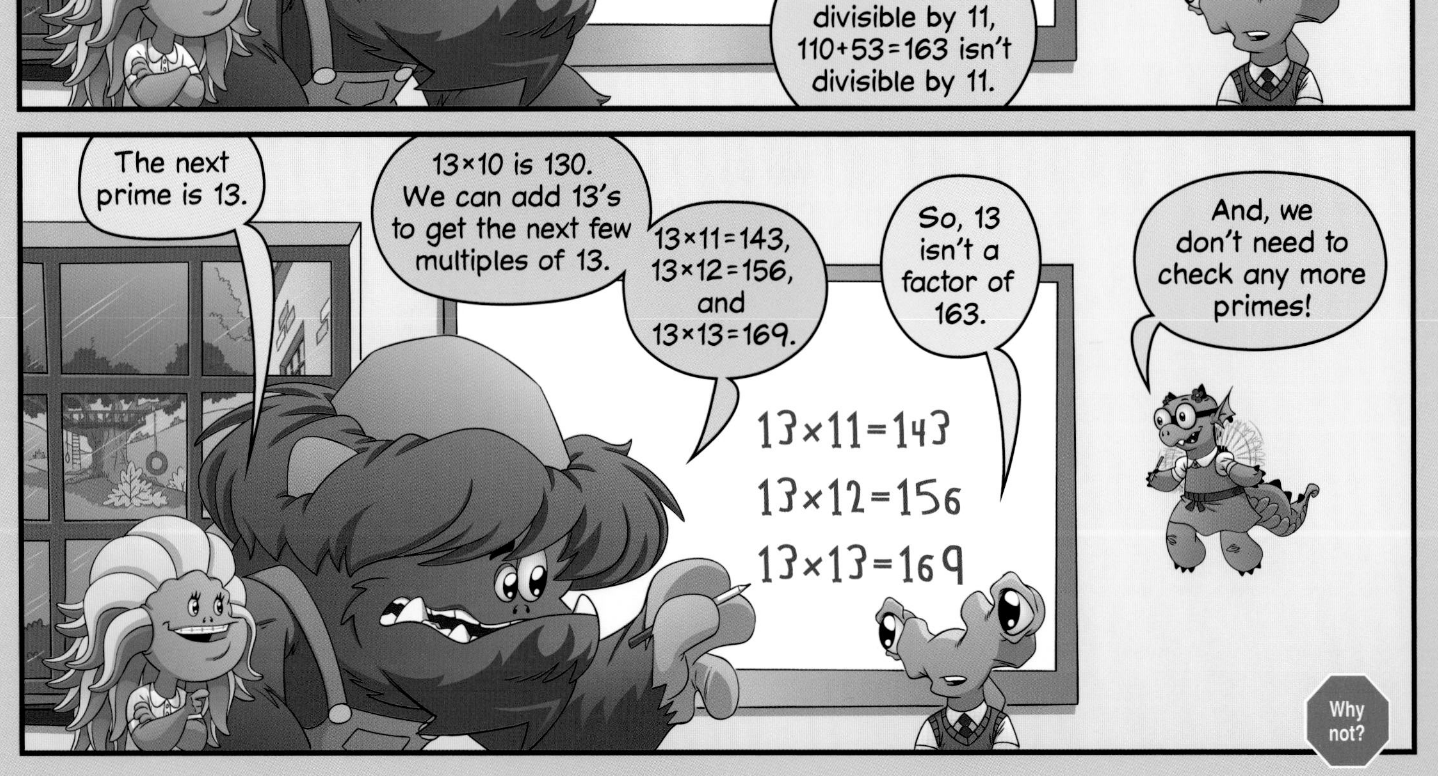
The next prime is 13.
13×10 is 130. We can add 13's to get the next few multiples of 13.
13×11=143, 13×12=156, and 13×13=169.
13×11=143
13×12=156
13×13=169
So, 13 isn't a factor of 163.
And, we don't need to check any more primes!
Why not?

Practice: Pages 15-37

Contents: Chapter 8

See page 38 in the Practice book for a recommended reading/practice sequence for Chapter 8.

Review 50
Can you remember what you've learned?

Adding Fractions 56
What is $\frac{3}{7}+\frac{2}{7}$?

Mixed Numbers 58
If 5 cans hold 16 ounces of Fun-Doh, how many ounces are in each can?

Rummy $\frac{7}{56}$ 63
How could you use a 5, a 6, and an 8 to create a meld equal to $\frac{1}{7}$?

Adding Mixed Numbers 64
How do we add $5\frac{7}{8}+4\frac{3}{8}$?

Story Problem 67
Is this what Ms. Q. had in mind?

Fraction Subtraction 68
How do you subtract $8\frac{8}{9}$ from $15\frac{4}{9}$?

Mental Math 73
What strategies do you use to compute with fractions?

Chapter 8: Fractions

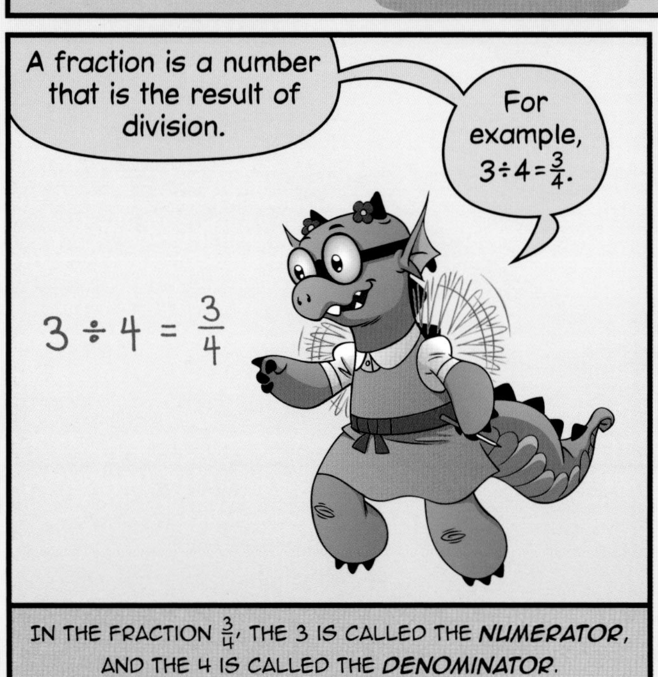

IN THE FRACTION $\frac{3}{4}$, THE 3 IS CALLED THE ***NUMERATOR***, AND THE 4 IS CALLED THE ***DENOMINATOR***.

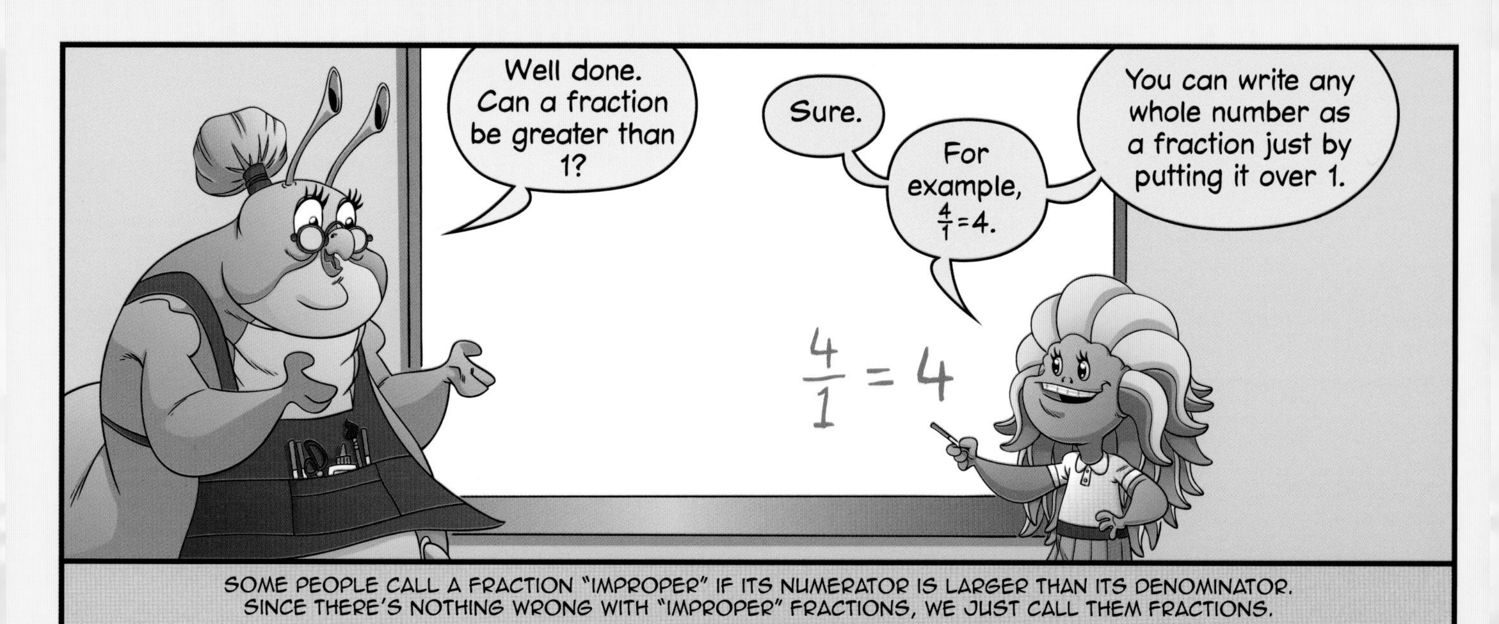

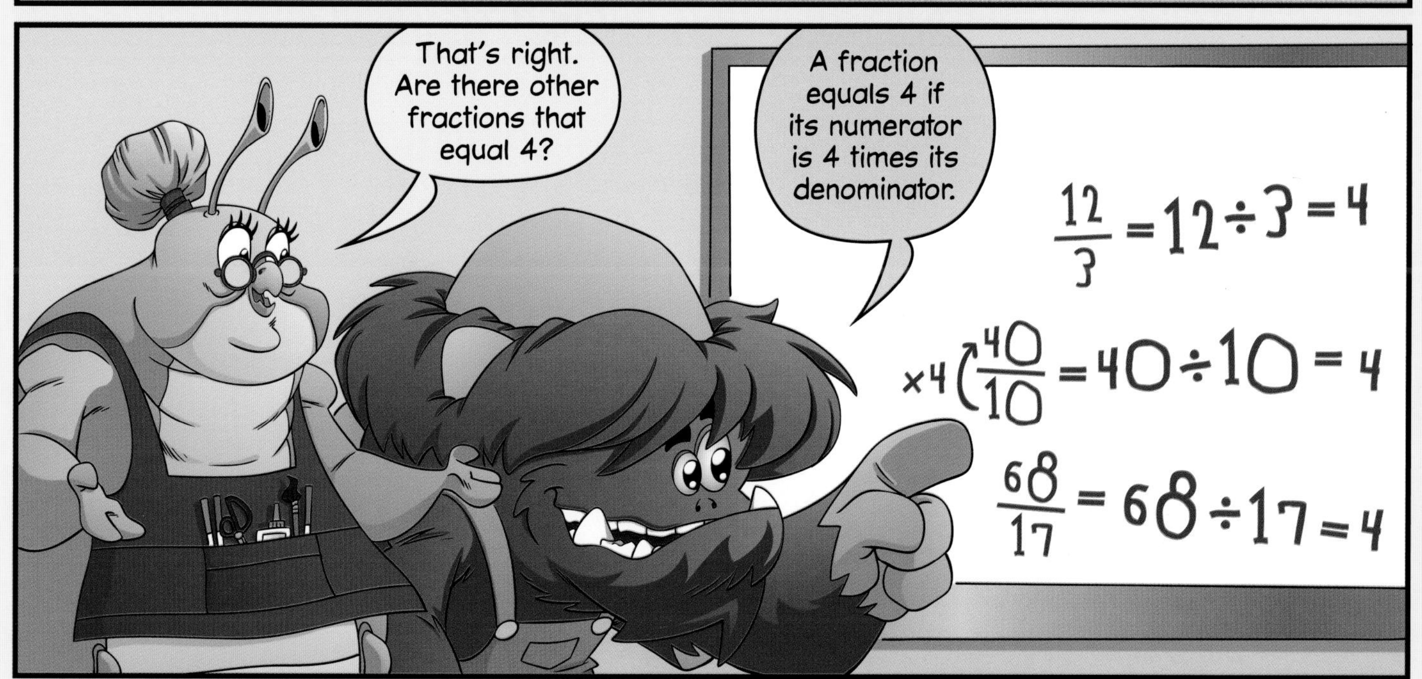

Is $\frac{28}{6}$ between 4 and 5?

We know that $\frac{24}{6} = 24 \div 6 = 4$.
And $\frac{30}{6} = 30 \div 6 = 5$.
So, $\frac{28}{6}$ is between 4 and 5.
4
5
$\frac{24}{6}$
$\frac{30}{6}$

$\frac{28}{6}$ is four sixths more than 4.
We can write $\frac{28}{6}$ as a ***mixed number:*** $4\frac{4}{6}$.
4
$4\frac{1}{6}$ $4\frac{2}{6}$ $4\frac{3}{6}$ $4\frac{4}{6}$ $4\frac{5}{6}$
5
$\frac{24}{6}$ $\frac{25}{6}$ $\frac{26}{6}$ $\frac{27}{6}$ $\frac{28}{6}$ $\frac{29}{6}$ $\frac{30}{6}$
$\frac{28}{6} = 4 + \frac{4}{6} = 4\frac{4}{6}$
A MIXED NUMBER CAN BE USED TO WRITE A FRACTION THAT IS GREATER THAN 1. A MIXED NUMBER IS WRITTEN AS A WHOLE NUMBER FOLLOWED BY A FRACTION THAT IS LESS THAN 1. LEARN MORE ABOUT MIXED NUMBERS BEGINNING ON PAGE 58.

We can simplify the fractional part of $4\frac{4}{6}$.
$4\frac{4}{6}$
How do we simplify $4\frac{4}{6}$?

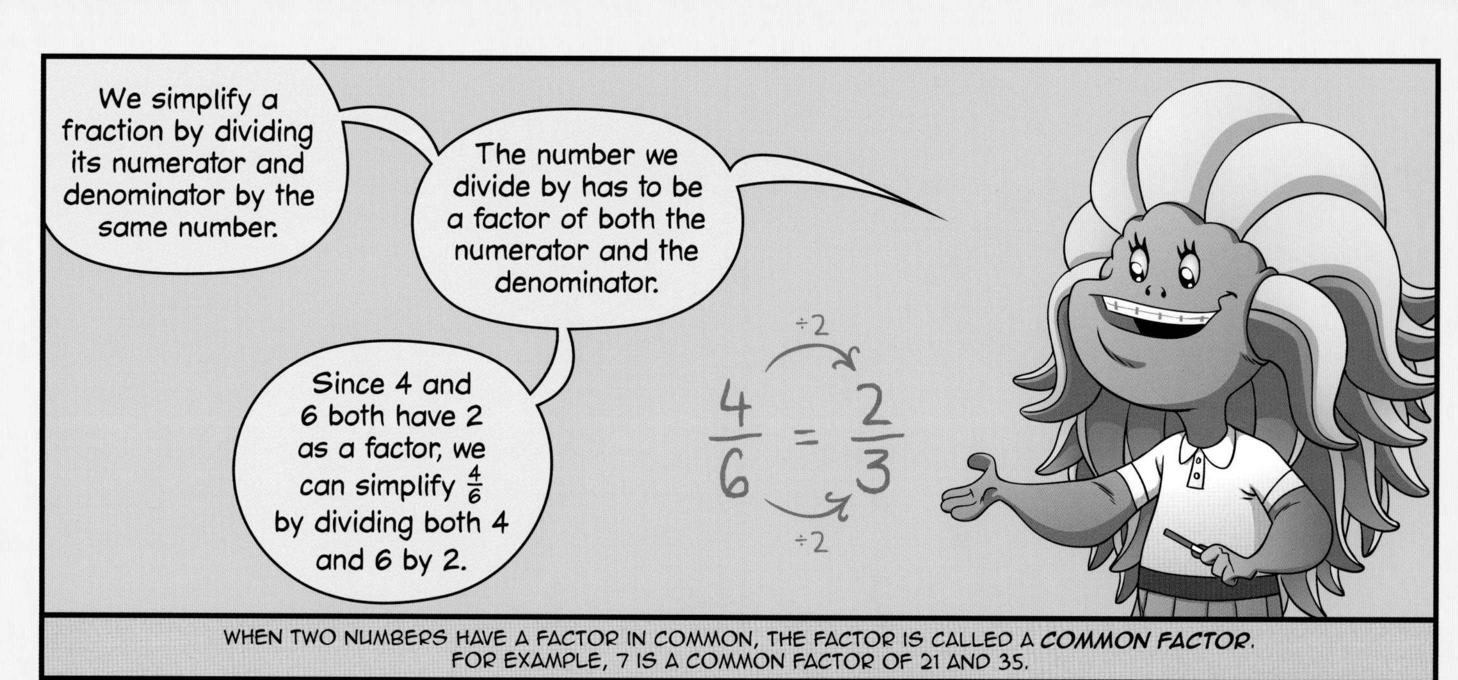
We simplify a fraction by dividing its numerator and denominator by the same number.
The number we divide by has to be a factor of both the numerator and the denominator.
Since 4 and 6 both have 2 as a factor, we can simplify $\frac{4}{6}$ by dividing both 4 and 6 by 2.
$\frac{4}{6} = \frac{2}{3}$
÷2
÷2
WHEN TWO NUMBERS HAVE A FACTOR IN COMMON, THE FACTOR IS CALLED A ***COMMON FACTOR***. FOR EXAMPLE, 7 IS A COMMON FACTOR OF 21 AND 35.

So, $4\frac{4}{6}$ simplifies to $4\frac{2}{3}$.
$4\frac{4}{6} = 4\frac{2}{3}$
Since the numerator and denominator don't have any common factors besides 1, we can't simplify any more.
When a fraction can't be simplified, we say that the fraction is in ***simplest form.***

Nice work. There's just one thing left to review.
<, >, or =
1. $\frac{4}{9}$ $\frac{5}{9}$
2. $\frac{7}{10}$ $\frac{7}{11}$
3. $\frac{8}{21}$ $\frac{3}{7}$
How do we compare the fractions in each of these three pairs?
Try it.

Five ninths is bigger than four ninths, because there are more ninths.
$\frac{4}{9} < \frac{5}{9}$

Seven tenths is bigger than seven elevenths, because tenths are bigger than elevenths.
$\frac{7}{10} > \frac{7}{11}$

It's harder to compare $\frac{8}{21}$ and $\frac{3}{7}$, because the two fractions don't have the same denominator or the same numerator.
But, we can ***convert*** one or both fractions to make them easier to compare.
How?
$\frac{8}{21}$ $\frac{3}{7}$

Converting a fraction is like simplifying, only in reverse.
To convert a fraction, we ***multiply*** its numerator and denominator by the same number.
For example, we can convert $\frac{3}{7}$ to $\frac{6}{14}$...
...or to $\frac{30}{70}$.
$\frac{3}{7} = \frac{6}{14}$ (×2, ×2)
$\frac{3}{7} = \frac{30}{70}$ (×10, ×10)

How can converting $\frac{3}{7}$ help us compare it to $\frac{8}{21}$?

IF YOU HAVE TROUBLE UNDERSTANDING THE MATERIAL IN THIS SECTION, WE RECOMMEND REVIEWING THE FRACTION CHAPTER IN BEAST ACADEMY 3D.

MATH TEAM
Adding Fractions
What do you get when you add $\frac{3}{7}+\frac{2}{7}$?
Isn't adding sevenths just like adding anything else?
What do you mean, Grogg?

Well, if I add 3 books and 2 books, I get 5 books.

If I add 3 pencils and 2 pencils, I get 5 pencils.

And if I add 3 sausage biscuits and 2 sausage biscuits, I get 5 sausage biscuits.

So, adding 3 sevenths plus 2 sevenths equals 5 sevenths.
$\frac{3}{7}+\frac{2}{7}=\frac{5}{7}$

We can show the addition on the number line.

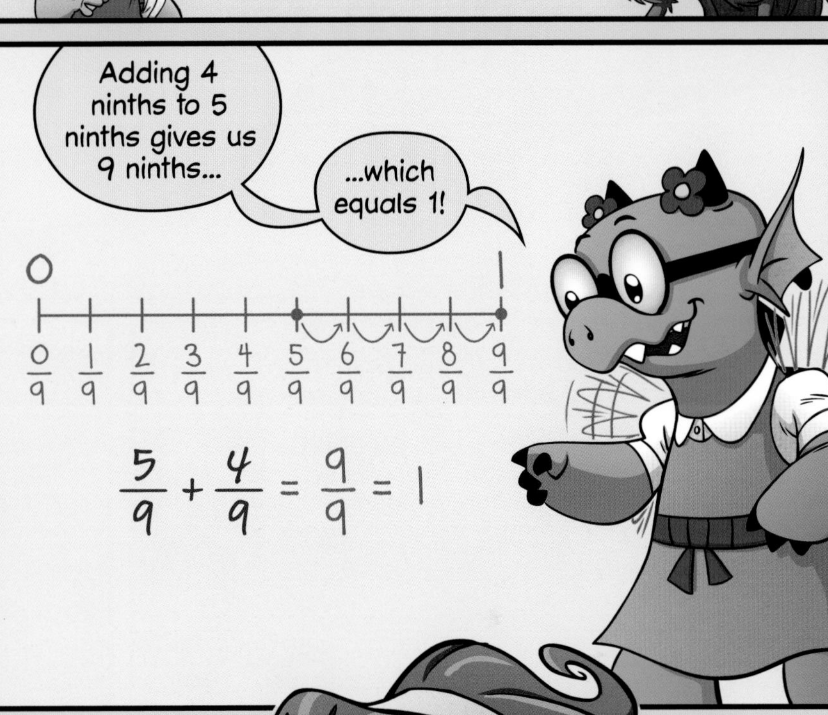

Nice work. Adding sevenths or ninths is just like adding books, or pencils, or sausage biscuits.

To add two fractions that have the same denominator...

...add the numerators, and keep the denominator.

$$\frac{a}{c}+\frac{b}{c}=\frac{a+b}{c}$$

Practice: Pages 39-45

How much does a can of Fun-Doh weigh?

Each can holds the same amount. So, the last ounce is divided equally among the 5 cans.
Each can has an additional $1 \div 5 = \frac{1}{5}$ ounce.
So, there are $3 + \frac{1}{5} = 3\frac{1}{5}$ ounces of Fun-Doh in each can?

Yup.
We can write the result of division as a fraction or a mixed number instead of a quotient and a remainder.
CLEARANCE
2-Sleeved Shirts
6-Legged Pants
4-Legged
Fun-Doh
I'm gonna need to see another example.

Take a look at this can of soup.

It holds 26 ounces, and says it will serve 3 monsters. How many ounces are in each serving?

I need to divide the number of ounces by the number of servings...
...26÷3.
Right, but since 26 is not divisible by 3, we write the number of ounces as a fraction or a mixed number.
How would you write 26÷3 as a fraction and as a mixed number?

SOCKS • SOCKS • SOCKS
I see. We can just write 26÷3 as a fraction.
$26 \div 3 = \frac{26}{3}$.
That's right.
So, each serving is $\frac{26}{3}$ ounces.
ADJUSTABLE TAIL HOLE

How much is $\frac{26}{3}$ ounces?
scratch scratch

It'll be easier to tell if we convert $\frac{26}{3}$ to a mixed number.

We know that $\frac{24}{3} = 8$.
$\frac{26}{3}$ is just $\frac{24}{3} + \frac{2}{3}$.
So, $\frac{26}{3}$ ounces is $8 + \frac{2}{3} = 8\frac{2}{3}$ ounces.
8
$8\frac{1}{3}$
$8\frac{2}{3}$
9
$\frac{24}{3}$
$\frac{25}{3}$
$\frac{26}{3}$
$\frac{27}{3}$

8
3) 26
−24
2
That makes sense.
If I divide 26 ounces of soup into 3 equal servings, each serving gets 8 whole ounces.
That leaves 2 ounces to split 3 ways.

So, each serving gets another $2 \div 3 = \frac{2}{3}$ ounce.
That makes $8 + \frac{2}{3} = 8\frac{2}{3}$ ounces of soup in each serving.

That's great!
To convert $\frac{26}{3}$ into a mixed number, we can use the quotient and remainder of 26÷3!
The quotient gives us the whole number part, and the remainder gives us the numerator of the fraction.
The denominator is the same as in the original fraction. $\frac{26}{3}=8\frac{2}{3}$.
Try $\frac{80}{7}$.
8
3) 26
−24
2
$8\frac{2}{3}$

80÷7 has quotient 11 and remainder 3.
So, $\frac{80}{7}=11\frac{3}{7}$.
11
7) 80
−77
3
$\frac{80}{7}=11\frac{3}{7}$

Now we know how to change a fraction into a mixed number.
How do we change a mixed number back to a fraction?
I don't know, but I bet we can figure it out.
Let's try $3\frac{4}{5}$.
How would you write $3\frac{4}{5}$ as a fraction?

3 $3\frac{1}{5}$ $3\frac{2}{5}$ $3\frac{3}{5}$ $3\frac{4}{5}$ 4
$\frac{15}{5}$ $\frac{16}{5}$ $\frac{17}{5}$ $\frac{18}{5}$ $\frac{19}{5}$ $\frac{20}{5}$
$3\frac{4}{5}$ is $3+\frac{4}{5}$.
And $3=\frac{15}{5}$.
So, $3\frac{4}{5}=\frac{15}{5}+\frac{4}{5}=\frac{19}{5}$.
Sunglasses

I could use some new sunglasses.
BEAST QUARTERLY

BEAST QUARTERLY

BEAST QUARTERLY

Store Hours
They never carry my style.

Rummy $\frac{7}{56}$

Rummy 7/56 is a 2-player card game similar to the popular card game Rummy 500. In Rummy 7/56, players score points by creating pairs or sets of equivalent fractions and mixed numbers, called melds.

The game uses a standard deck of playing cards with the face cards (K, Q, J) removed. Aces are treated as 1's. Shuffle and deal 9 cards face down to both players. Place the remaining cards face down in a pile called the stock. Flip the top card from the stock to begin the discard pile. During play, discards are placed as shown on the right so that each card in the pile is visible.

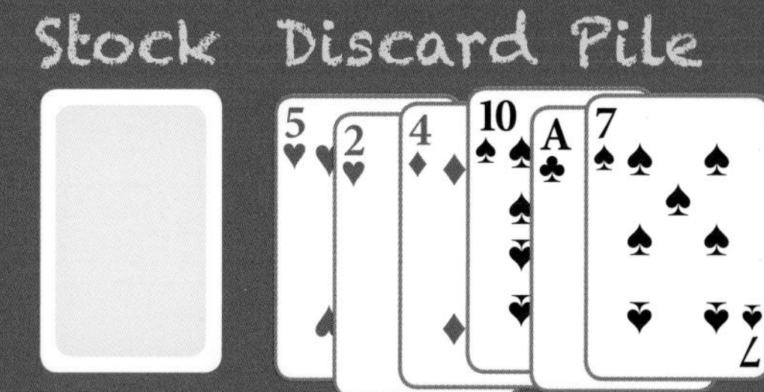

Play

Players alternate turns. During each turn, a player draws, plays melds (optional), then discards.

Drawing

A player has two options:

a) Draw the top card from the stock; or

b) Take any card from the discard pile and all cards on top of it. The bottom card drawn from the discard pile must be played in a meld on the same turn it is drawn.

For example, if Winnie chooses to take the 4 of diamonds from the discard pile above, she must also pick up the three cards above it (10♠, A♣, and 7♠). Winnie must then play a meld with the 4 of diamonds. The other cards drawn may be played in a meld or kept in her hand.

Playing Melds

A player may play one or more melds face up. A meld is 2 or more cards from the player's hand representing digits of a fraction or mixed number. Each meld played must equal a meld that has already been played (by either player), or another meld played at the same time. Two sample melds are shown on the right.

- Duplicate fractions may not be used (8/9=8/9 is not allowed).
- Fractions that equal whole numbers are not allowed (for example, 24/6 and 8/2).
- Players cannot play all their cards but must keep at least one card to discard.

Melds

9♠ / 7♥ $\left(\frac{9}{7} = 1\frac{2}{7}\right)$ A♠ 2♦ / 7♣

8♠ / 5♦ 6♣ $\left(\frac{8}{56} = \frac{1}{7}\right)$ A♥ / 7♦

Discarding

To end his or her turn, a player must discard one card to the top of the discard pile.

The End and Scoring

The game ends when a player has no cards after discarding, or when a player's turn ends and the stock is empty. Players score one point for each card they have played in a meld and lose one point for each card that remains in their hand. The player with the most points wins.

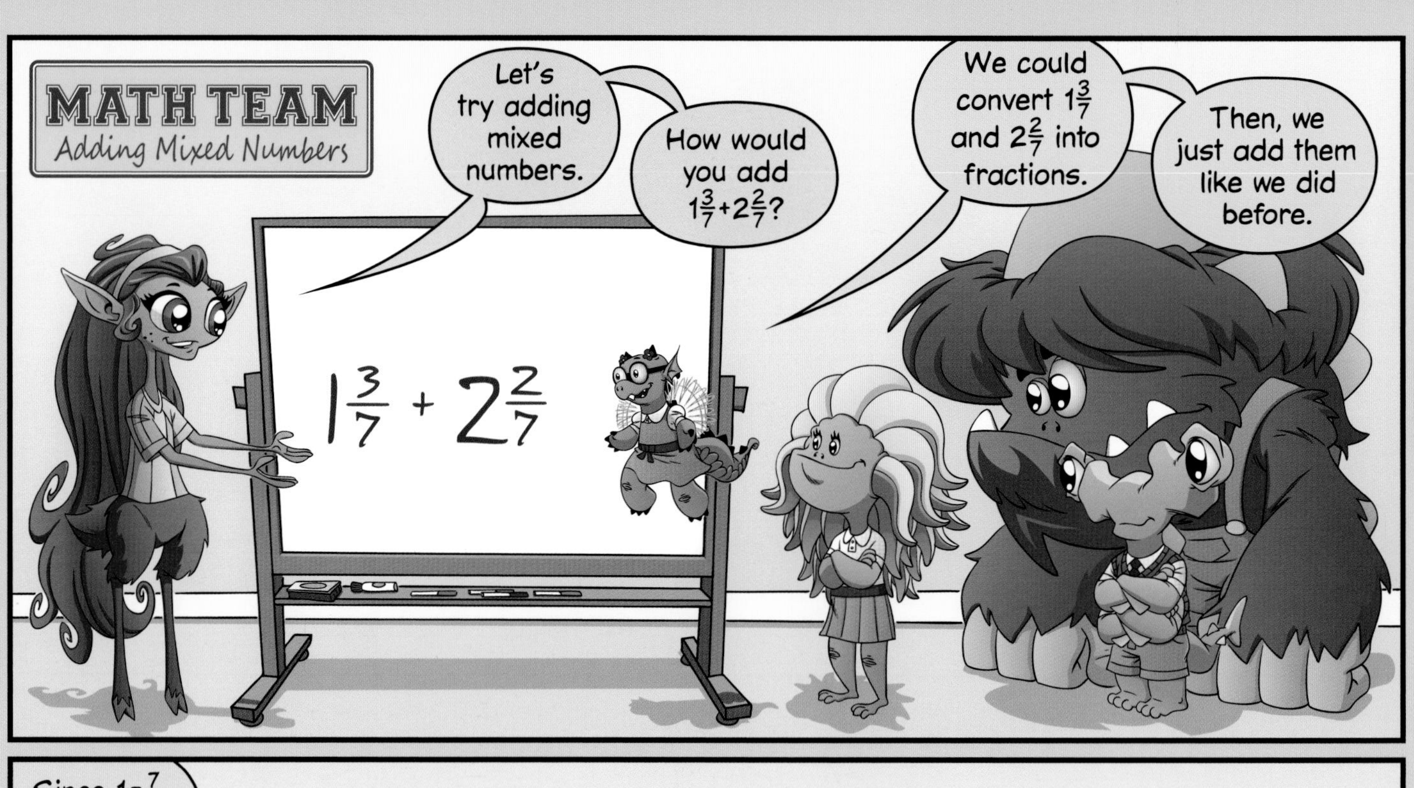
MATH TEAM
Adding Mixed Numbers
Let's try adding mixed numbers.
How would you add $1\frac{3}{7}+2\frac{2}{7}$?
We could convert $1\frac{3}{7}$ and $2\frac{2}{7}$ into fractions.
Then, we just add them like we did before.
$1\frac{3}{7}+2\frac{2}{7}$

$1\frac{3}{7}+2\frac{2}{7}=\frac{10}{7}+\frac{16}{7}=\frac{26}{7}=3\frac{5}{7}$
Since $1=\frac{7}{7}$, we have $1\frac{3}{7}=\frac{7}{7}+\frac{3}{7}=\frac{10}{7}$.
And since $2=\frac{14}{7}$, we have $2\frac{2}{7}=\frac{14}{7}+\frac{2}{7}=\frac{16}{7}$.
Adding $\frac{10}{7}+\frac{16}{7}$...
...we get $\frac{26}{7}$.
26÷7 has quotient 3 and remainder 5. So, we can write $\frac{26}{7}$ as a mixed number: $3\frac{5}{7}$.

Can't we add mixed numbers without converting them to fractions?
You can.
How would you add $1\frac{3}{7}+2\frac{2}{7}$ ***without*** converting?
Try it.

We know that $1\frac{3}{7}=1+\frac{3}{7}$...
...and $2\frac{2}{7}=2+\frac{2}{7}$.
We can add the whole numbers and the fractions separately.
$1\frac{3}{7}+2\frac{2}{7}$
$=(1+\frac{3}{7})+(2+\frac{2}{7})$
$=(1+2)+(\frac{3}{7}+\frac{2}{7})$
$=3+\frac{5}{7}$
$=3\frac{5}{7}$
$1+2=3$.
$\frac{3}{7}+\frac{2}{7}=\frac{5}{7}$.
So, $1\frac{3}{7}+2\frac{2}{7}=3\frac{5}{7}$.

Great job. It's usually a lot easier to add mixed numbers by adding the whole numbers and the fractions separately.
Sometimes things get a little tricky, though.
Try this one.
$5\frac{7}{8}+4\frac{3}{8}$
Try it.

5+4 is 9, and $\frac{7}{8}+\frac{3}{8}=\frac{10}{8}$.
So, $5\frac{7}{8}+4\frac{3}{8}=9\frac{10}{8}$!
$5\frac{7}{8}+4\frac{3}{8}$
$=5+4+\frac{7}{8}+\frac{3}{8}$
$=9\frac{10}{8}$

I don't think we're done.
The fractional part of a mixed number has to be ***less*** than 1.
And, we can simplify $\frac{10}{8}$.
$=9\frac{10}{8}$
How would you write $9\frac{10}{8}$ as a mixed number in simplest form?

WE COULD HAVE ALSO CONVERTED $\frac{10}{8}$ TO $1\frac{2}{8}$ FIRST, THEN SIMPLIFIED $1\frac{2}{8}$ TO $1\frac{1}{4}$.

Practice: Pages 46-55

name: Grogg

Homework: Write and solve a story problem involving fractions in the space below.

Fraction Jackson

1/2 Monster, 1/2 amazing!

It was a quarter past 7. Fraction Jackson was half asleep when he got the call...

Exactly 5 1/4 seconds later...

Jackson knew exactly what needed to be done and got to work right away.

With the fraction problem solved, the city was safe again. . . Or was it?

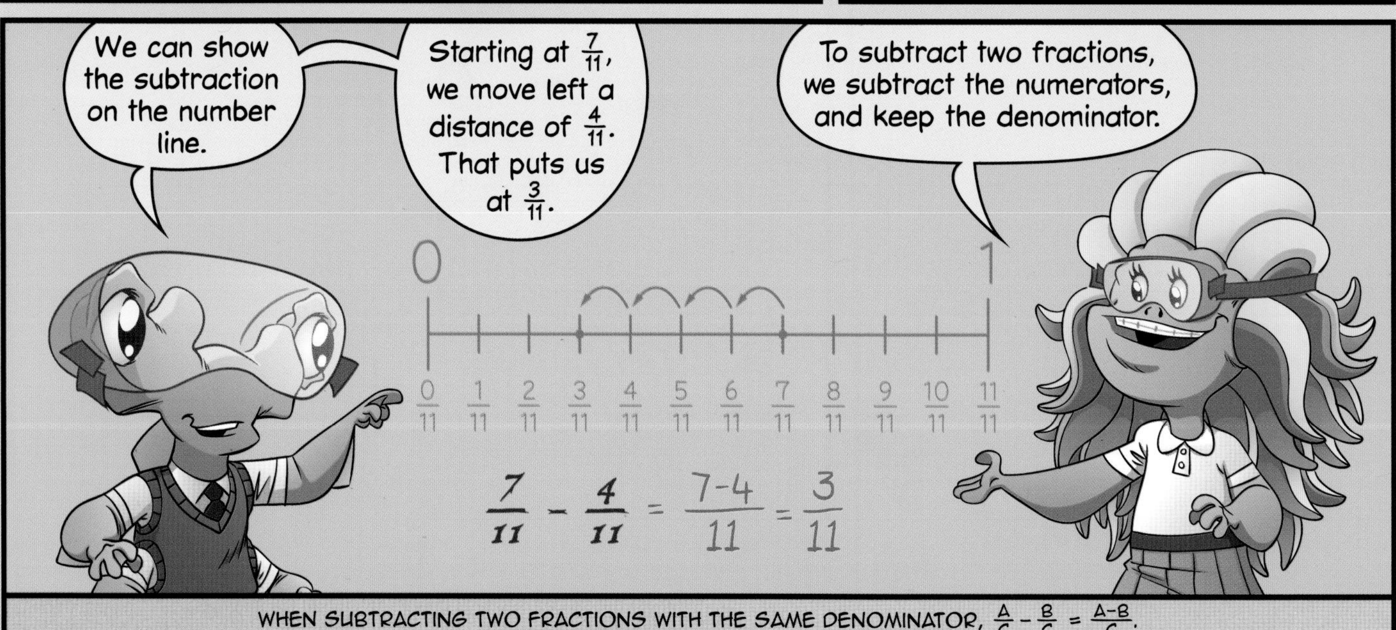

WHEN SUBTRACTING TWO FRACTIONS WITH THE SAME DENOMINATOR, $\frac{A}{C} - \frac{B}{C} = \frac{A-B}{C}$.

Aye. Let's try includin' some mixed numbers. What do you get when you subtract $2\frac{3}{5} - 1\frac{1}{5}$?
We could convert $2\frac{3}{5}$ and $1\frac{1}{5}$ into fractions.
$2\frac{3}{5} - 1\frac{1}{5}$

$2\frac{3}{5} = \frac{13}{5}$, and $1\frac{1}{5} = \frac{6}{5}$.
So, $2\frac{3}{5} - 1\frac{1}{5} = \frac{13}{5} - \frac{6}{5}$.
And $\frac{13}{5} - \frac{6}{5}$ equals $\frac{7}{5}$, which is $1\frac{2}{5}$.
$2\frac{3}{5} - 1\frac{1}{5}$
$= \frac{13}{5} - \frac{6}{5}$
$= \frac{7}{5}$
$= 1\frac{2}{5}$

If we stack the mixed numbers, we can subtract without converting.
$2\frac{3}{5}$
$-1\frac{1}{5}$

2-1=1...
...and $\frac{3}{5} - \frac{1}{5} = \frac{2}{5}$.
$2\frac{3}{5}$
$-1\frac{1}{5}$
$1\frac{2}{5}$

Good thinkin'! Subtractin' two mixed numbers can get tricky, though. Try this one.
$15\frac{4}{9} - 8\frac{8}{9}$
Let's try stacking the subtraction again!
Try it.

WRITING $15\frac{4}{9}$ AS $14\frac{13}{9}$ IS CALLED ***REGROUPING*** THE MIXED NUMBER.

How long will a $12\frac{1}{4}$-inch dowel be after cutting $3\frac{3}{4}$ inches from one end?

We can write $12\frac{1}{4}$ as $11\frac{5}{4}$, then subtract.
$11-3=8$, and $\frac{5}{4}-\frac{3}{4}=\frac{2}{4}$, so...
...$11\frac{5}{4}-3\frac{3}{4}=8\frac{2}{4}$.
And, we can simplify $\frac{2}{4}$ to $\frac{1}{2}$.
$11\frac{5}{4}$
$-3\frac{3}{4}$
$8\frac{2}{4}=8\frac{1}{2}$

So, the trimmed dowel will be $8\frac{1}{2}$ inches long.
Aye.
Once you've got yer $8\frac{1}{2}$-inch dowel, attach it to the stand we made yesterday.
What are we making, anyway.

Peg legs!

I'll grab some paints so we can decorate 'em!

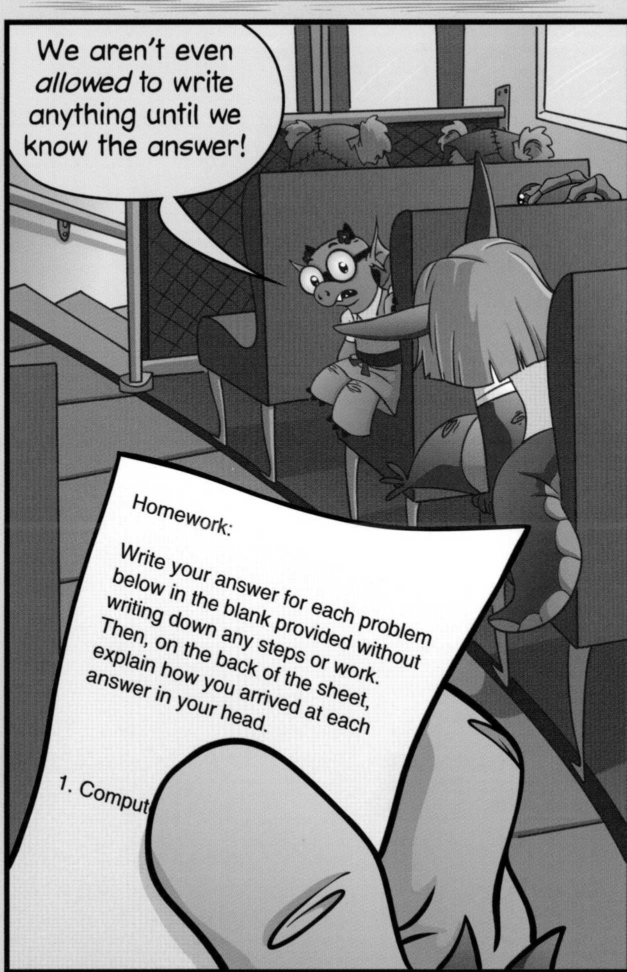

*$11\frac{14}{15}+\frac{8}{15} = 11+\frac{14}{15}+\frac{8}{15} = 11+\left(\frac{14}{15}+\frac{8}{15}\right)$.

To add eight fifteenths to $11\frac{14}{15}$, we can start by adding one fifteenth to $11\frac{14}{15}$. That gives us 12.
Then, we just add 7 more fifteenths.
$12+\frac{7}{15}=12\frac{7}{15}$.
IN MATH, WE WRITE:
$11\frac{14}{15}+\frac{8}{15}$
$=11\frac{14}{15}+\left(\frac{1}{15}+\frac{7}{15}\right)$
$=\left(11\frac{14}{15}+\frac{1}{15}\right)+\frac{7}{15}$
$=12+\frac{7}{15}$
$=12\frac{7}{15}$
BEAST ACADEMY

Great idea, Winnie.
$11\frac{14}{15}+\frac{8}{15}$ $=12\frac{7}{15}$.
BEAST

What's the second problem?
BEAS

$6\frac{3}{11}-3\frac{9}{11}$.
ADEMY

I got it.
Me, too.
I know!
Huh?
How'd you guys do that so fast?
BEAST ACADEMY
How could you compute $6\frac{3}{11}-3\frac{9}{11}$ in your head?

I found the difference by counting up.
From $3\frac{9}{11}$ to 4 is $\frac{2}{11}$.
Then, from 4 to $6\frac{3}{11}$ is $2\frac{3}{11}$ more.
So, the difference is $2\frac{3}{11}+\frac{2}{11}=2\frac{5}{11}$.
3 4 5 6 7
$\frac{2}{11}$
$2\frac{3}{11}$
$2\frac{5}{11}$
I got the same answer a different way.
To make the subtraction easier, I added $\frac{2}{11}$ to both numbers.
Adding $\frac{2}{11}$ to both numbers in a subtraction problem doesn't change their difference.
$6\frac{3}{11}-3\frac{9}{11}$ equals $6\frac{5}{11}-4$, which is $2\frac{5}{11}$.
$6\frac{3}{11}-3\frac{9}{11}$
3 4 5 6 7
$6\frac{5}{11}-4=2\frac{5}{11}$
I did something else.
$3\frac{9}{11}$ is close to 4, and subtracting 4 is a lot easier than subtracting $3\frac{9}{11}$.
$3\frac{9}{11}$ is $\frac{2}{11}$ less than 4.
$6\frac{3}{11}-3\frac{9}{11}$
$=6\frac{3}{11}-4+\frac{2}{11}$
$=2\frac{3}{11}+\frac{2}{11}$
$=2\frac{5}{11}$
So, to take away $3\frac{9}{11}$, I took away 4, then gave back $\frac{2}{11}$.
There are lots of different ways to think about subtraction. We all got $2\frac{5}{11}$!
What's the last problem?
We need to add $\frac{1}{13}+\frac{2}{13}+\frac{3}{13}+\frac{4}{13}+\frac{5}{13}+\frac{6}{13}+\frac{7}{13}+\frac{8}{13}+\frac{9}{13}+\frac{10}{13}+\frac{11}{13}+\frac{12}{13}$.
Can you find an easy way to solve this one in your head?

3. Compute: $\frac{1}{13}+\frac{2}{13}+\frac{3}{13}+\frac{4}{13}+\frac{5}{13}+\frac{6}{13}+\frac{7}{13}+\frac{8}{13}+\frac{9}{13}+\frac{10}{13}+\frac{11}{13}+\frac{12}{13}$

Practice: Pages 56-71

Contents: Chapter 9

See page 72 in the Practice book for a recommended reading/practice sequence for Chapter 9.

Negatives 80
What temperature is 15 degrees colder than 8 degrees?

Less Than Zero 82
Which is greater: -7 or -15?

Adding Integers 87
How do we add -89+104?

Grogg's Notes 93
What makes an integer an integer?

Subtracting Integers 94
How do we subtract -3 from 5?

Rearranging 100
Why is it sometimes useful to write subtraction as addition?

Lizzie's Notes 109
Can you write an expression that means the same thing as -a-(-b) without using any negatives?

Chapter 9: Integers

R&G
Negatives
Brrrrr... It's cold out.
The forecast said the temperature is supposed to drop 15 degrees overnight.
It's only 8 degrees now. How can the temperature drop 15 degrees?
We can't take 15 from 8!

The temperature is going to fall **below** zero.
Huh?
Here, look at this thermometer.
Sometimes it gets even colder than 0 degrees!

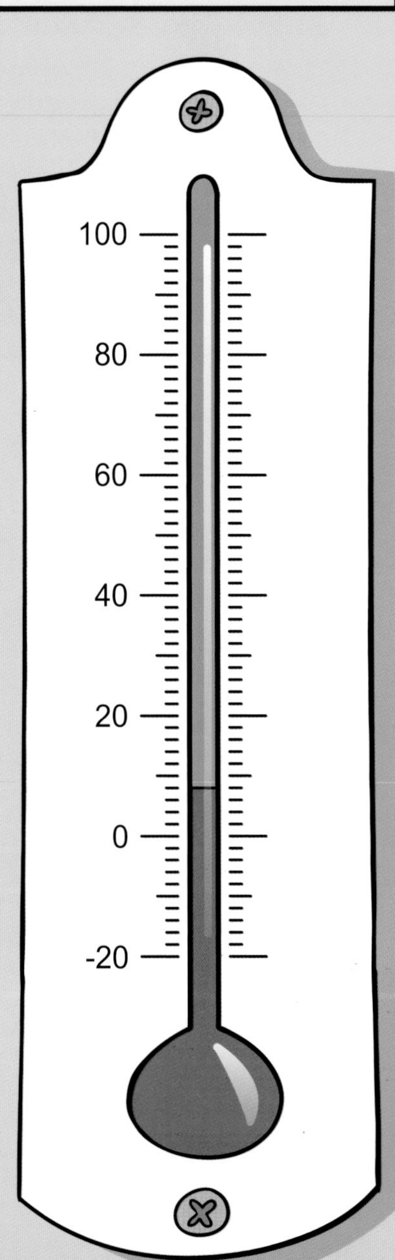
100
80
60
40
20
0
-20

When the temperature falls below zero, we get **negative** temperatures!
It's supposed to get down to negative 7 degrees tonight!
Negative 7?

We use negative numbers to describe temperatures that are below 0.
Negative 7 degrees means 7 degrees below zero.
We write the number negative 7 like this:

Interesting. So, if it's -7 degrees out, and the temperature drops 1 degree...
...the temperature will be -8 degrees?
Yep.
Cool.
You mean cold!

Cold is right! Lets get back indoors before our tail freezes to the ground again.

THE NUMBER −4 IS READ "NEGATIVE FOUR," −3 IS READ "NEGATIVE THREE," AND SO ON.

-4 -3 -2 -1 0 1 2 3 4
The negative symbol lets you know that -3 is less than zero, and the 3 tells you how far it is from zero.
-3 is three units less than zero.
Prealgebra
So, the number -100 is 100 units less than 0...
...and -200 is 200 units less than 0.
-200 -100 0 100
I guess that means that -200 is ***less*** than -100, right?
That's right, Grogg.
Ordering numbers can be tricky when negatives are included.
How would you order these five numbers from least to greatest?
0 -7 9 14 -15
Try it.

NUMBERS TO THE RIGHT OF ZERO ON THE NUMBER LINE ARE CALLED ***POSITIVE*** NUMBERS.

When we compare two ***negative*** numbers...
...the one that is farther from zero is less than the one that is closer to zero.
Right again, Grogg.
-15
-7
0
9
14
-15<-7
Does anyone know the term we use to describe a number's distance from zero?

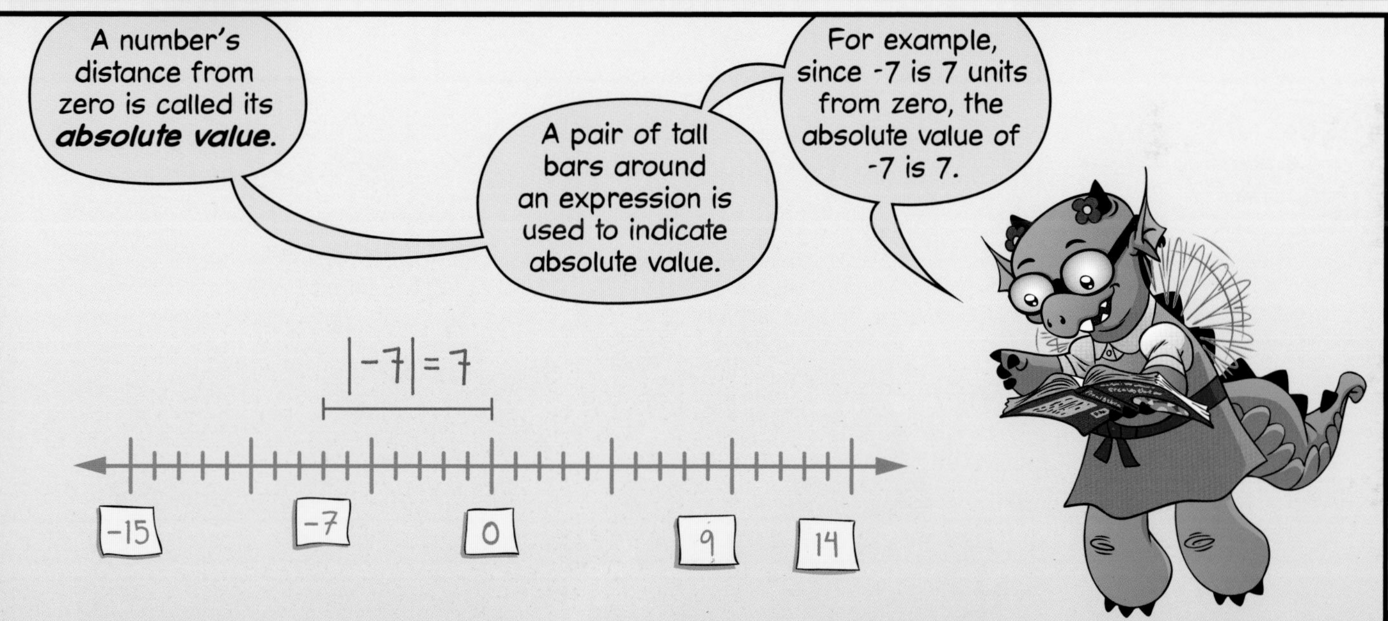
A number's distance from zero is called its ***absolute value***.
A pair of tall bars around an expression is used to indicate absolute value.
For example, since -7 is 7 units from zero, the absolute value of -7 is 7.
|-7| = 7
-15
-7
0
9
14

Very good, Lizzie!
Now, how do we compare each of these pairs of expressions?
<, >, or =
-1 -2
|-24| 23
|6| |-6|
Try it.

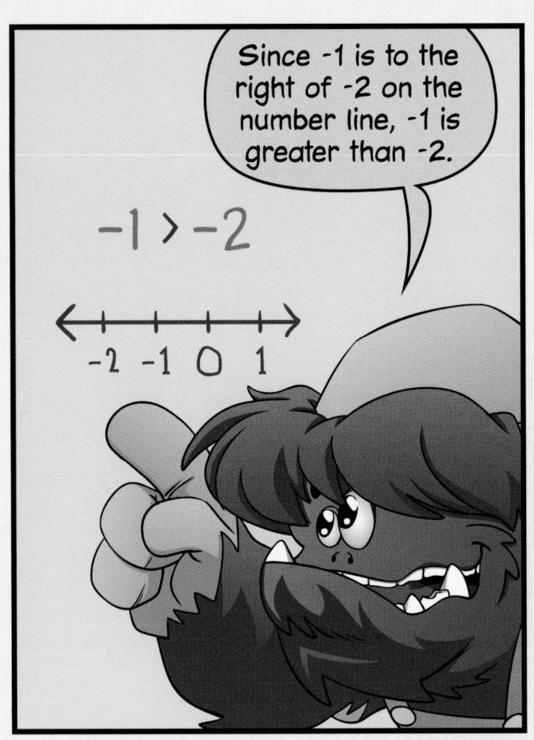
Since -1 is to the right of -2 on the number line, -1 is greater than -2.
-1 > -2
-2 -1 0 1

Since -24 is 24 units from zero, its absolute value is 24.
24 is greater than 23.
So, the absolute value of -24 is greater than 23.
|-24| = 24
|-24| > 23

6 and -6 are both 6 units from zero, but on opposite sides of zero.
So, the absolute values of 6 and -6 are equal.
6 = 6
|6| = |-6|

6 and -6 are called ***opposites***.
What a coincidence. Today is Opposite Day!
Don't you mean today is ***not*** Opposite Day?
TWO INTEGERS ARE OPPOSITES IF THEY ARE THE SAME DISTANCE FROM ZERO, BUT ON OPPOSITE SIDES OF ZERO.

What happened to Grogg?
Ms. Q. broke his brain again.

MATH TEAM
Adding Integers
I have good news and bad news.
One of the Bots has a virus and we were unable to schedule a math meet this month.
Cool. What's the bad news?

The *good* news is that you have been invited to participate in the Math Relays event at the end of the school year.
Only the top teams in the area are selected to compete each year.
Grogg! That *was* the bad news.

Is that Max kid going to be there?
Probably, but he'll have to bring a team.
This is one contest he can't win on his own.
We're great as a team.
It'll be good to have a chance to compete against him again.

Enough about competitions, let's learn how to add integers.
Integers?
A number that does not have a fractional part is called an ***integer***.

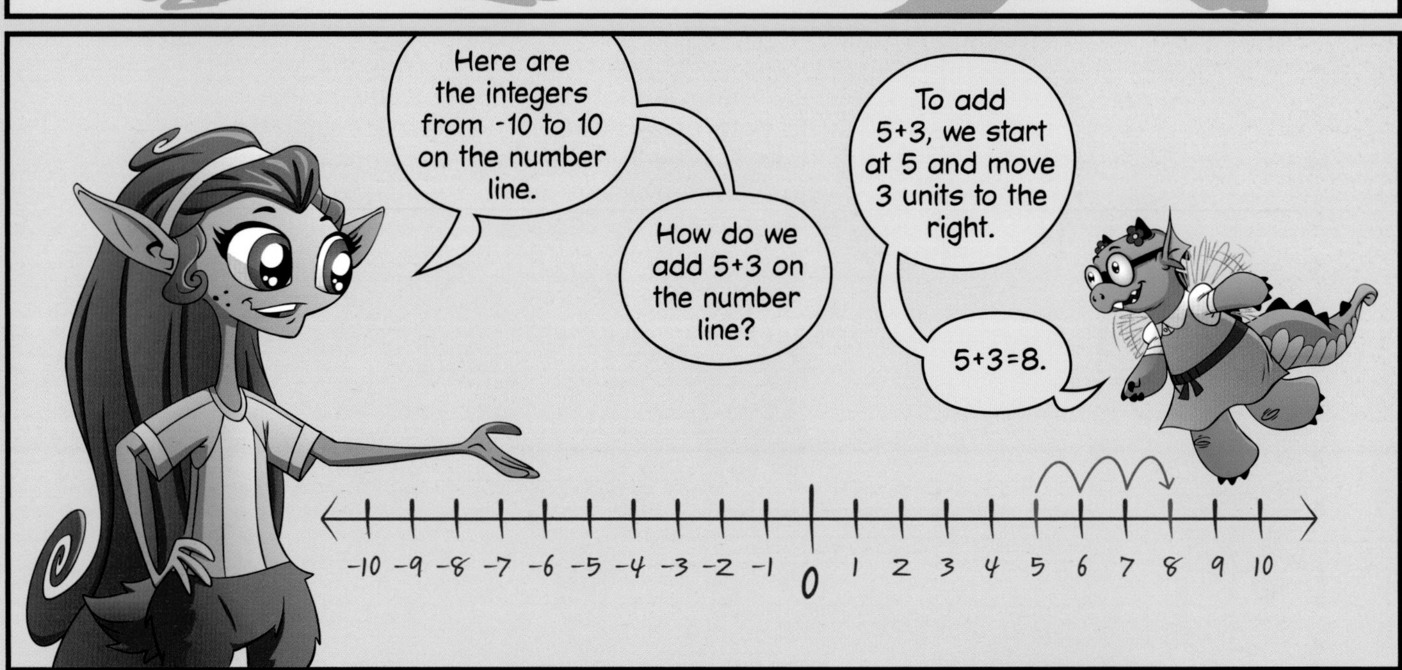
Here are the integers from -10 to 10 on the number line.
How do we add 5+3 on the number line?
To add 5+3, we start at 5 and move 3 units to the right.
5+3=8.
-10 -9 -8 -7 -6 -5 -4 -3 -2 -1 0 1 2 3 4 5 6 7 8 9 10

To add 3 to any number on the number line, we start at the number and move 3 spaces to the right.
Exactly.
So, what do we get when we add -5+3?

Try it.

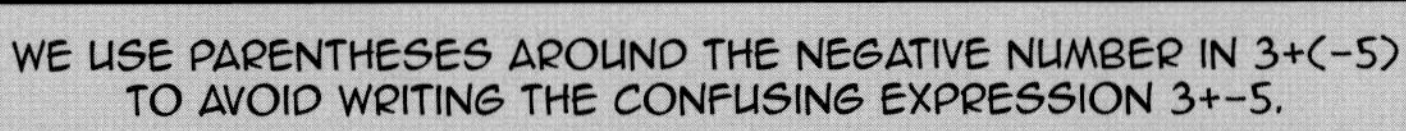
WE USE PARENTHESES AROUND THE NEGATIVE NUMBER IN 3+(-5) TO AVOID WRITING THE CONFUSING EXPRESSION 3+-5.

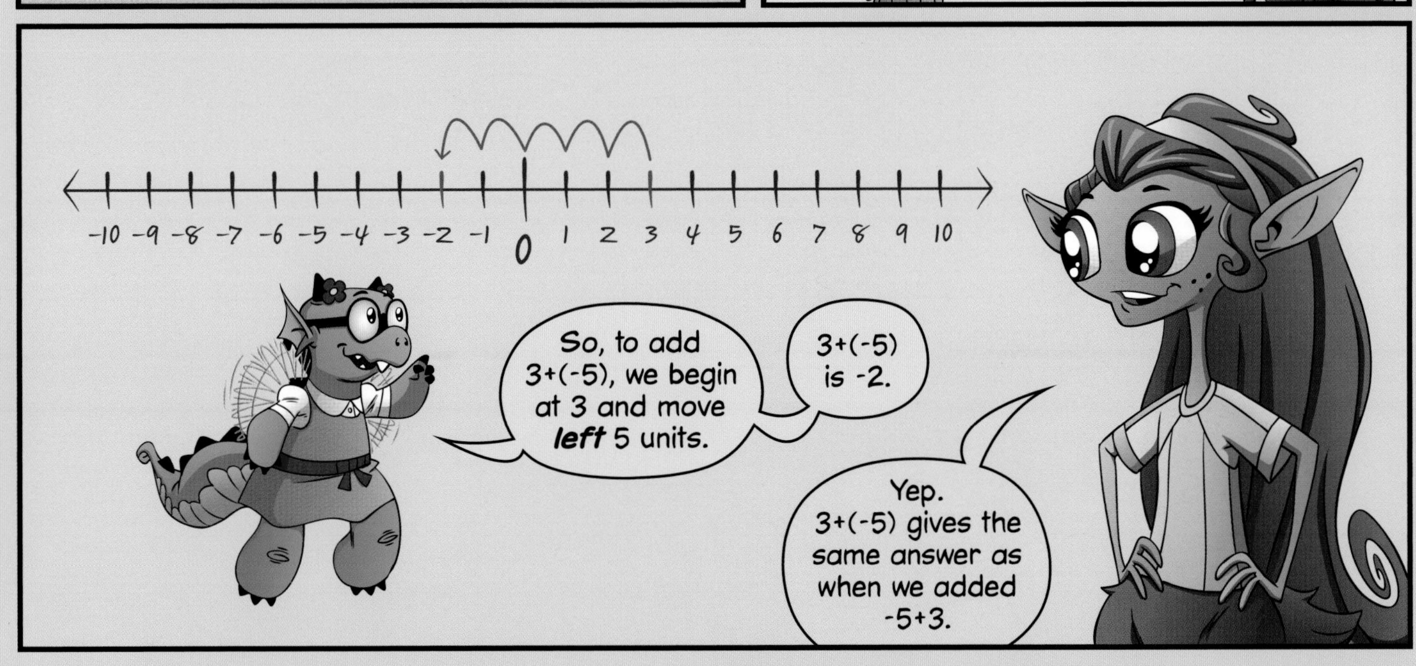

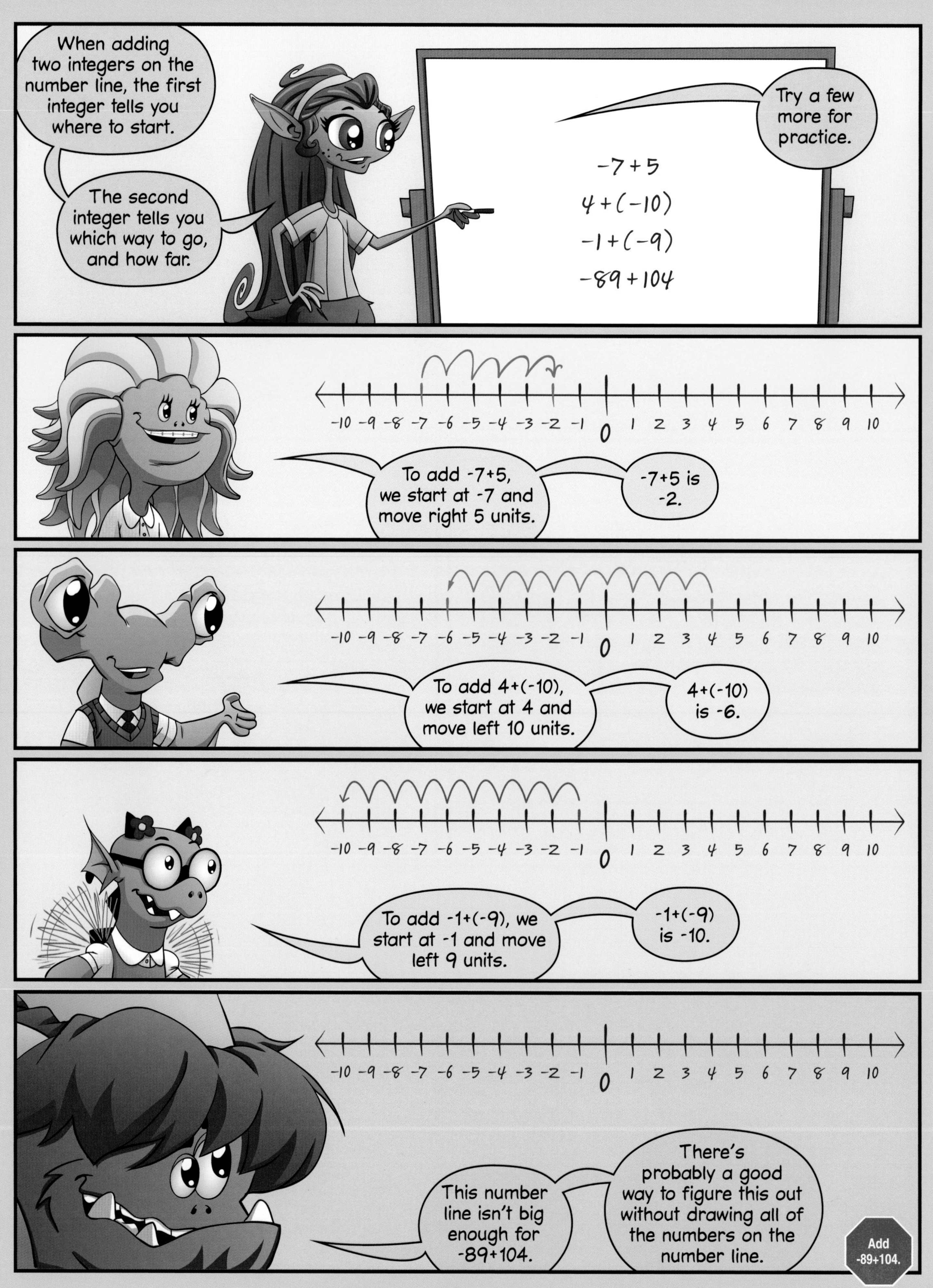
When adding two integers on the number line, the first integer tells you where to start.
The second integer tells you which way to go, and how far.
Try a few more for practice.
-7+5
4+(-10)
-1+(-9)
-89+104
-10 -9 -8 -7 -6 -5 -4 -3 -2 -1 0 1 2 3 4 5 6 7 8 9 10
To add -7+5, we start at -7 and move right 5 units.
-7+5 is -2.
-10 -9 -8 -7 -6 -5 -4 -3 -2 -1 0 1 2 3 4 5 6 7 8 9 10
To add 4+(-10), we start at 4 and move left 10 units.
4+(-10) is -6.
-10 -9 -8 -7 -6 -5 -4 -3 -2 -1 0 1 2 3 4 5 6 7 8 9 10
To add -1+(-9), we start at -1 and move left 9 units.
-1+(-9) is -10.
-10 -9 -8 -7 -6 -5 -4 -3 -2 -1 0 1 2 3 4 5 6 7 8 9 10
This number line isn't big enough for -89+104.
There's probably a good way to figure this out without drawing all of the numbers on the number line.
Add -89+104.

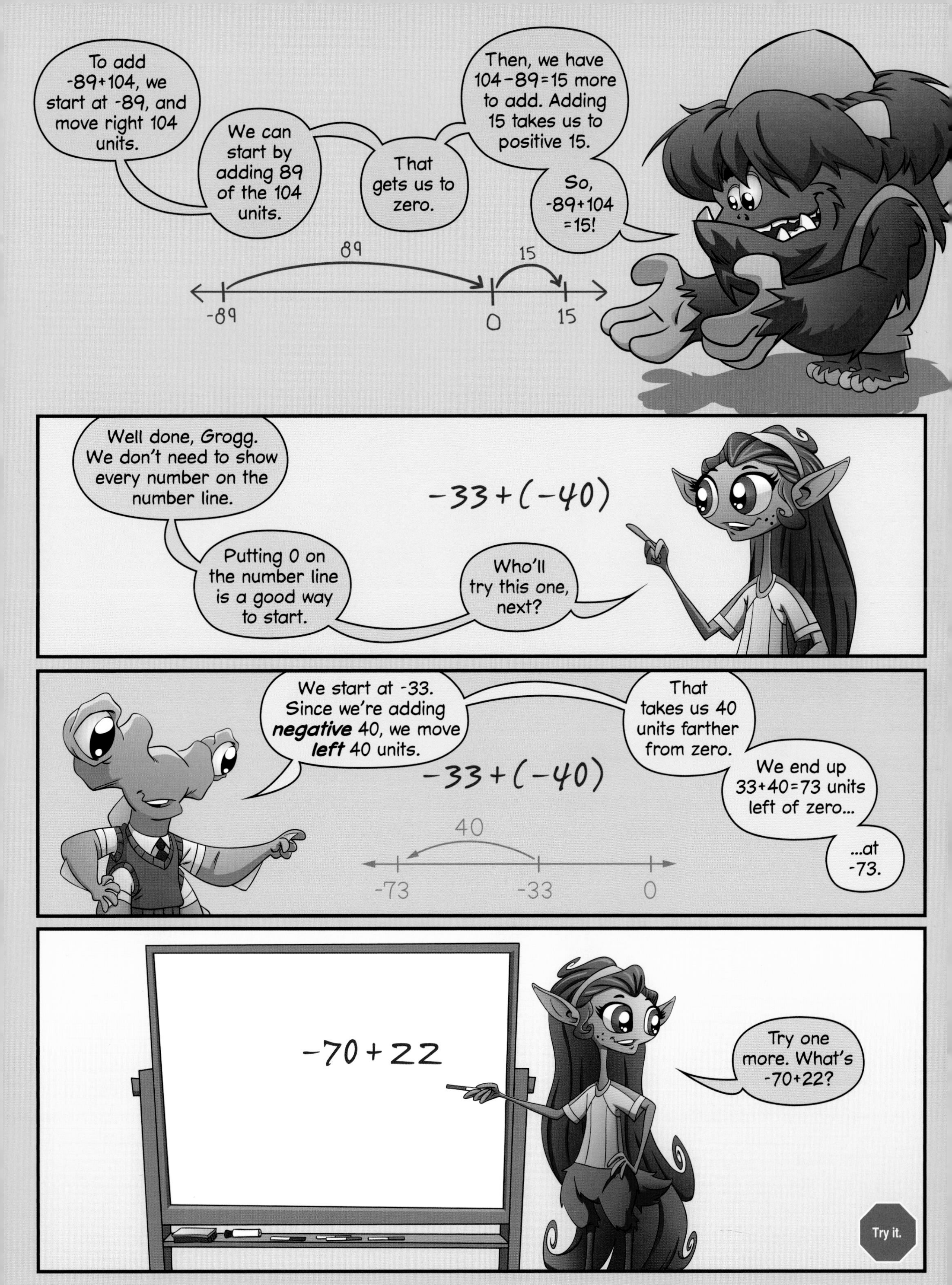
To add -89+104, we start at -89, and move right 104 units.
We can start by adding 89 of the 104 units.
That gets us to zero.
Then, we have 104−89=15 more to add. Adding 15 takes us to positive 15.
So, -89+104 =15!
89
15
-89
0
15
Well done, Grogg. We don't need to show every number on the number line.
Putting 0 on the number line is a good way to start.
Who'll try this one, next?
−33+(−40)
We start at -33. Since we're adding **negative** 40, we move **left** 40 units.
That takes us 40 units farther from zero.
We end up 33+40=73 units left of zero...
...at -73.
−33+(−40)
40
-73
-33
0
−70+22
Try one more. What's -70+22?
Try it.

Practice: Pages 73-89

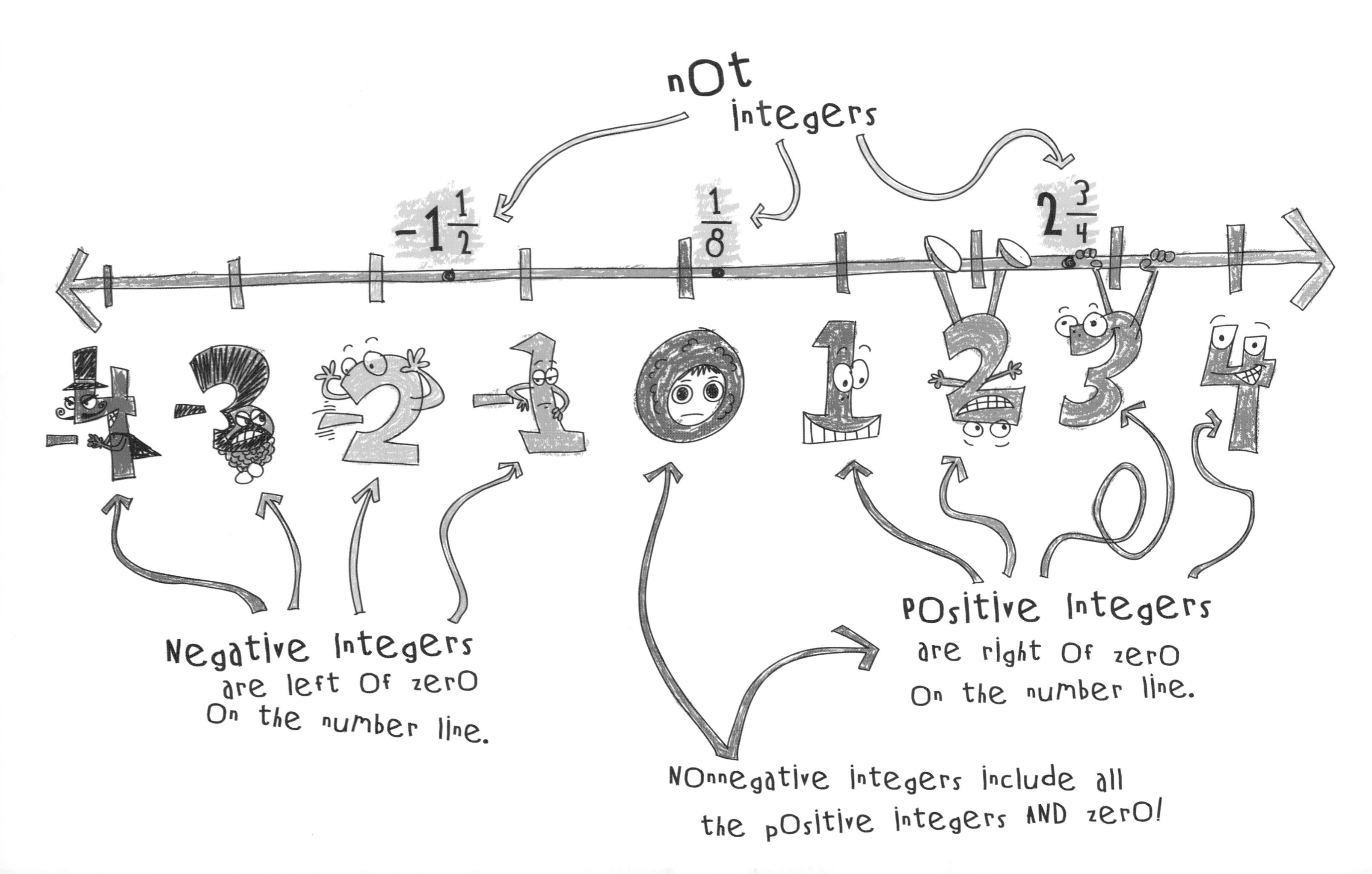

Integer: A number without a fractional part
not integers
$-1\frac{1}{2}$
$\frac{1}{8}$
$2\frac{3}{4}$
-4 -3 -2 -1 0 1 2 3 4
Negative Integers are left of zero on the number line.
Positive Integers are right of zero on the number line.
Nonnegative integers include all the positive integers AND zero!

WoodShop
SUBTRACTING INTEGERS
Captain Kraken?
Aye, lad.
How do we subtract 5−9?

Arrr... Let's try some examples 'n' see if we can figure it out.
Start with one ye' already know. What's 9−5?
That's easy. If I start with 9 and I take away 5, I have 4 left over. So, 9−5=4.
Be there other ways to think about subtraction?

We can use addition to solve a subtraction problem.
To answer 9−5, we can find the number we add to 5 to get 9.
Since 4+5=9, 9−5=4.
4 + 5 = 9
9 − 5 = 4

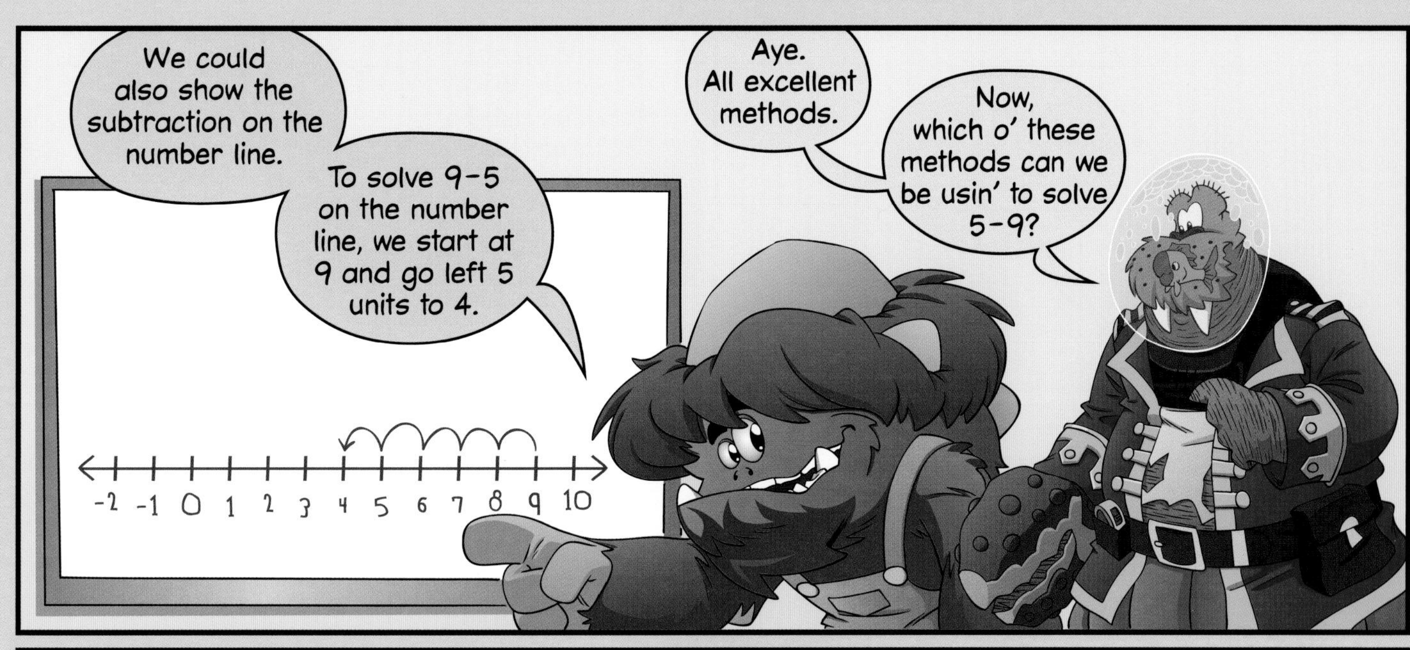
We could also show the subtraction on the number line.
To solve 9-5 on the number line, we start at 9 and go left 5 units to 4.
Aye. All excellent methods.
Now, which o' these methods can we be usin' to solve 5-9?
-2 -1 0 1 2 3 4 5 6 7 8 9 10

Grogg's number line method seems best. To subtract 5-9, we can start at 5 and move left 9 units.
We subtract 5 to get to zero, then 4 more to get to -4.
5-9=-4.
4
5
-5 -4 -3 -2 -1 0 1 2 3 4 5 6 7

My way is a little trickier, but it still works.
To answer 5-9, we can find the number we add to 9 to get 5.
Since (-4)+9=5, 5-9=-4.
-4+9=5
5-9=-4

Good.
Try a few more.
7-10
-6-13
-8-4
5-(-3)
Try them.

When we subtract 7−10, we go 3 units below 0. So, 7−10=-3.
7 - 10
=-3
3
7
-3 -2 -1 0 1 2 3 4 5 6 7
We can solve -6−13 on the number line, too.
Starting at -6, we go 13 units to the left to land on -19.
-6 - 13
= -19
13
-19
-6
0
-8−4 is -12.
-8 - 4
= -12
4
-12
-8
0
Hmmm. I'm gonna need a little help here.
How do we subtract a ***negative?***
5 - (-3)
? ? ?
0
5
Maybe we can try to solve 5-(-3) using addition.
To answer 5-(-3), we need to find the number we add to -3 to get 5.
5 - (-3) = ___
___ + (-3) = 5
Since 8+(-3)=5...
...5-(-3) must be 8!
5 - (-3) = 8
8 + (-3) = 5

*THIS IS SO IMPORTANT, WE WILL REPEAT IT:
SUBTRACTING A NUMBER IS THE SAME AS ADDING ITS OPPOSITE.

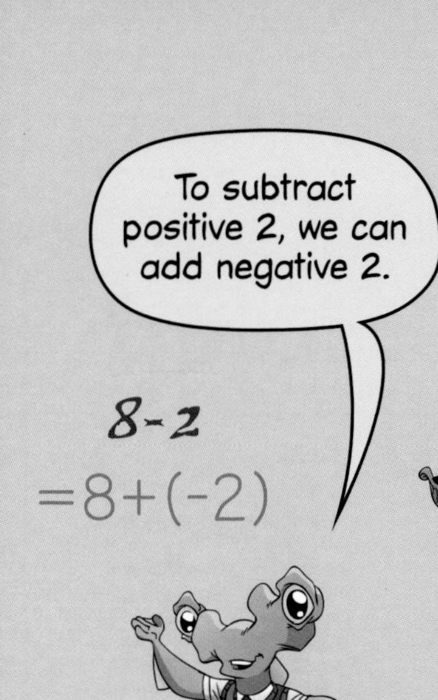

To subtract positive 2, we can add negative 2.
8-2
=8+(-2)

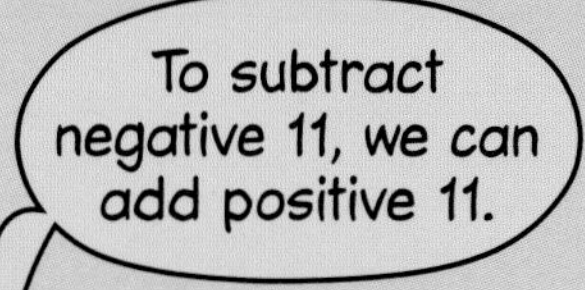

To subtract negative 11, we can add positive 11.

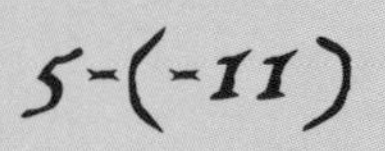

5-(-11)

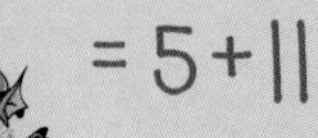

=5+11

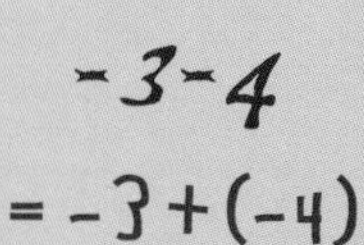

-3-4
=-3+(-4)
To subtract positive 4, we can add negative 4.
And to subtract negative 7, we can add positive 7.

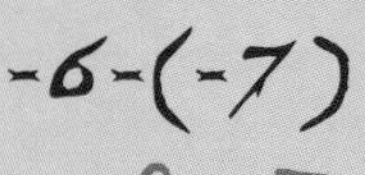

-6-(-7)

=-6+7

Beautiful work.

Now, compute the value of each expression usin' what you've learned about addition.

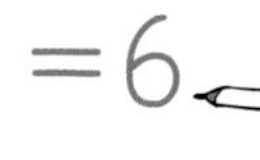

8-2
=8+(-2)
=6

5-(-11)
=5+11
=16

-3-4
=-3+(-4)
=-7

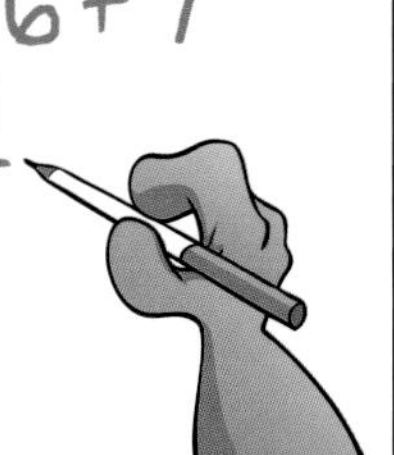

-6-(-7)
=-6+7
=1

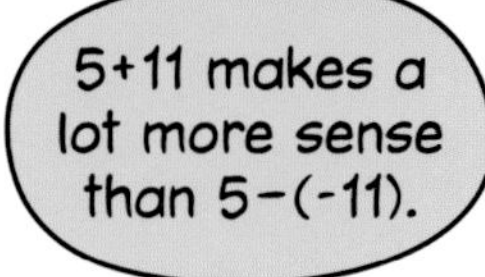

5+11 makes a lot more sense than 5-(-11).

5-(-11)
=5+11
=16

But to compute 8-2, it's better to just subtract.
8-2
=8+(-2)
=6

Aye, 'tis not always helpful to change subtraction to addition.
Let's try usin' some bigger numbers.
Find each o' these differences.
42-(-19)
-17-61

To subtract -19, we add its opposite.
42-(-19) equals 42+19=61.
42-(-19)
=42 + 19
=61

We can compute -17-61 with or without changing the subtraction to addition.
Either way, we start at -17 and move 61 units to the left.
-17-61
= -17+(-61)
61
-17
0

Arrr... Ye' be a smart bunch.
We land at -78.
-17-61
= -17+(-61)
61
-78
-17
0

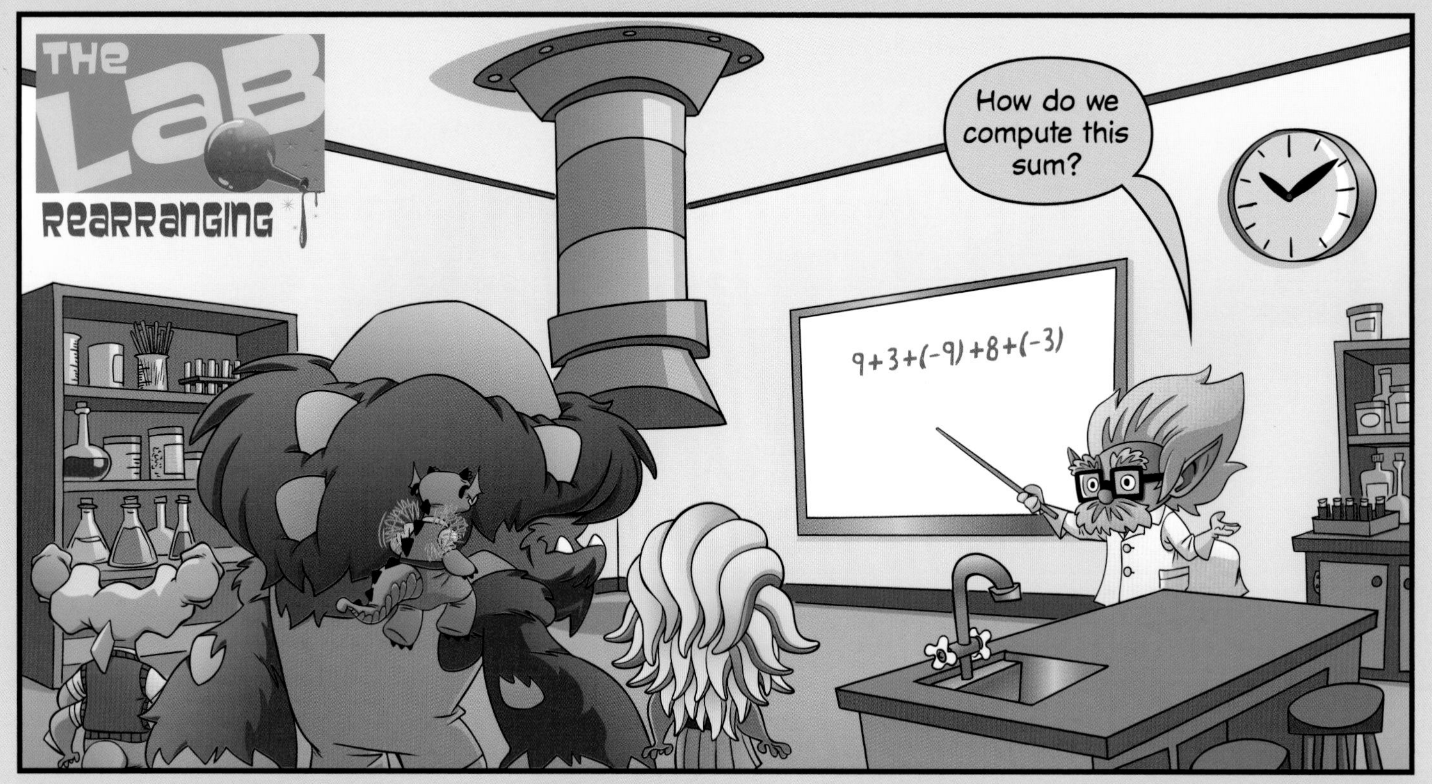

$9+3+(-9)+8+(-3)$

How can changing the order of the addition make the problem easier?

I see.
We can add the 9 to the -9...
...and the 3 to the -3.
9+(-9) and 3+(-3) both equal zero.
So, the sum of all five numbers is 0+0+8=8.
That *is* easier.
9+3+(-9)+8+(-3)
= 9+(-9)+3+(-3)+8
= 0 + 0 + 8
= 8
How would you compute the value of *this* expression?
99-100-99
Since addition is ***commutative*** and ***associative***, we can add as many terms as we like in any order that we like.
Very good.
THE ***COMMUTATIVE PROPERTY OF ADDITION*** STATES THAT $a+b=b+a$ FOR ALL VALUES OF a AND b.
THE ***ASSOCIATIVE PROPERTY OF ADDITION*** STATES THAT $(a+b)+c=a+(b+c)$ FOR ALL VALUES OF a, b, AND c.

We work from left to right. First, 99-100=-1.
Then, -1-99 =-100.
Very good.
99-100-99
= -1 - 99
= -100

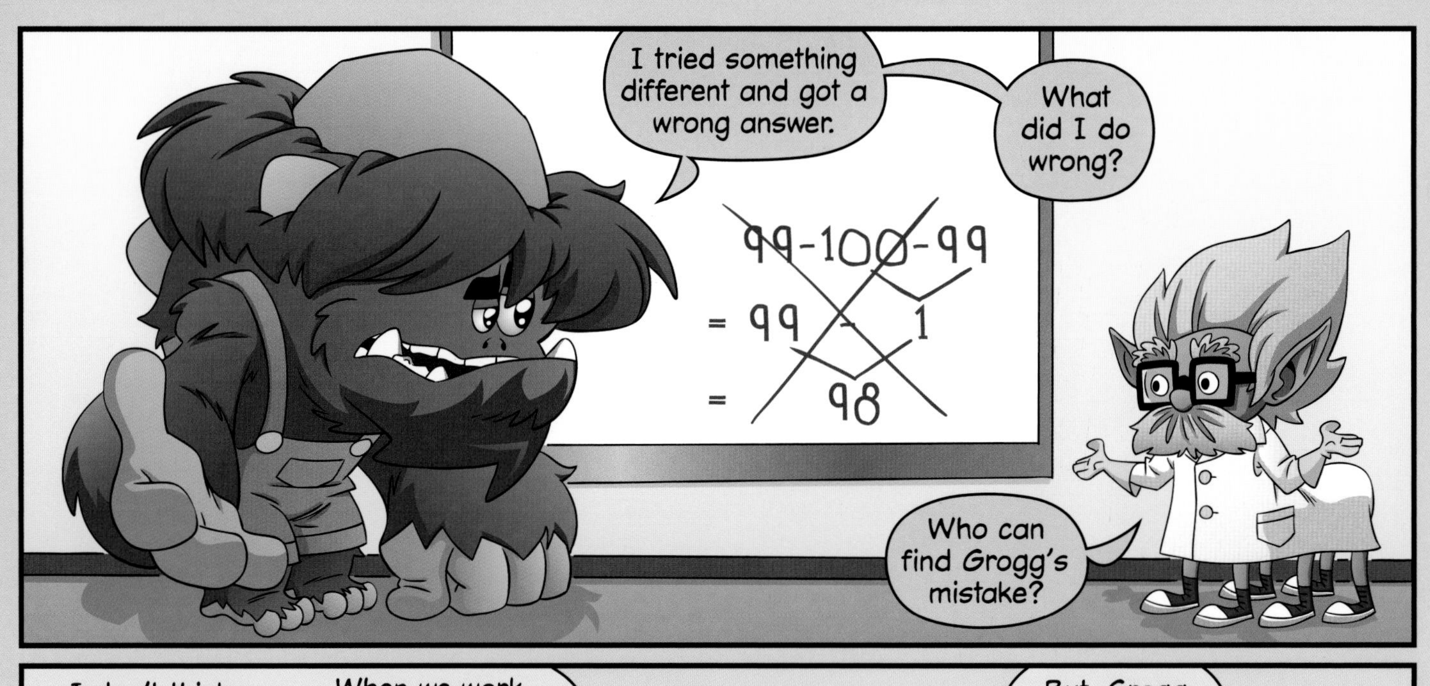
I tried something different and got a wrong answer.
What did I do wrong?
99-100-99
= 99 - 1
= 98
Who can find Grogg's mistake?

I don't think we can **subtract** integers in any order we want.
When we work from left to right, we subtract 99−100 first.
But, Grogg subtracted 100−99 first.
These expressions are not equal.
(99-100)-99
= -1 - 99
= -100
99-(100-99)
= 99 - 1
= 98

Very good. Subtraction is ***not*** associative.
$(a-b)-c$ is not always equal to $a-(b-c)$.
(99-100)-99
= -1 - 99
= -100
99-(100-99)
= 99 - 1
= 98
Is subtraction commutative?
Is $a-b$ always equal to $b-a$?

Nope. 3-2 is *not* equal to 2-3.
3-2=1
2-3=-1

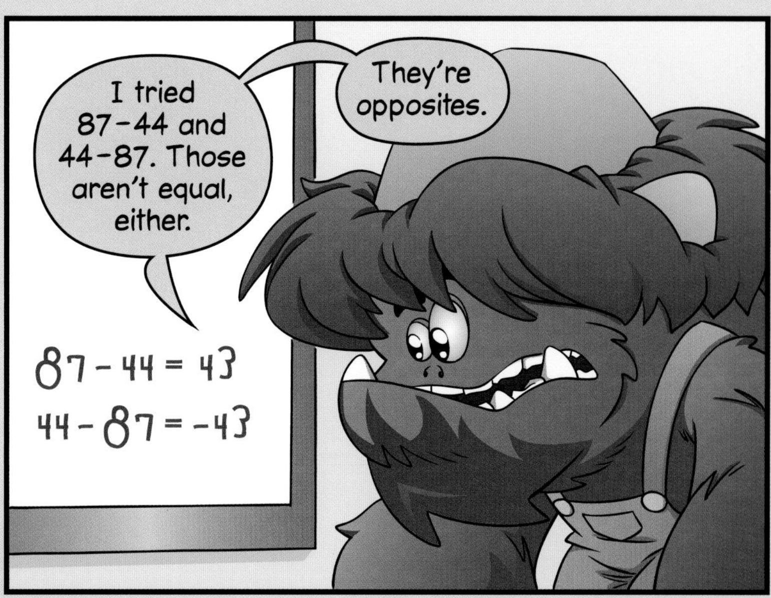
I tried 87-44 and 44-87. Those aren't equal, either.
They're opposites.
87-44=43
44-87=-43

So, subtraction is ***not*** commutative.
3-2=1
2-3=-1
So, we cannot change the order of the integers in a subtraction expression.
Right again. Subtraction is neither commutative nor associative.

I think there ***is*** a way to rearrange the terms in this expression.
99-100-99

What if we write the subtraction as addition?
99 - 100 - 99
=99+(-100)+(-99)
Great idea, Grogg.
Then, we can ***add*** the integers in any order we want.

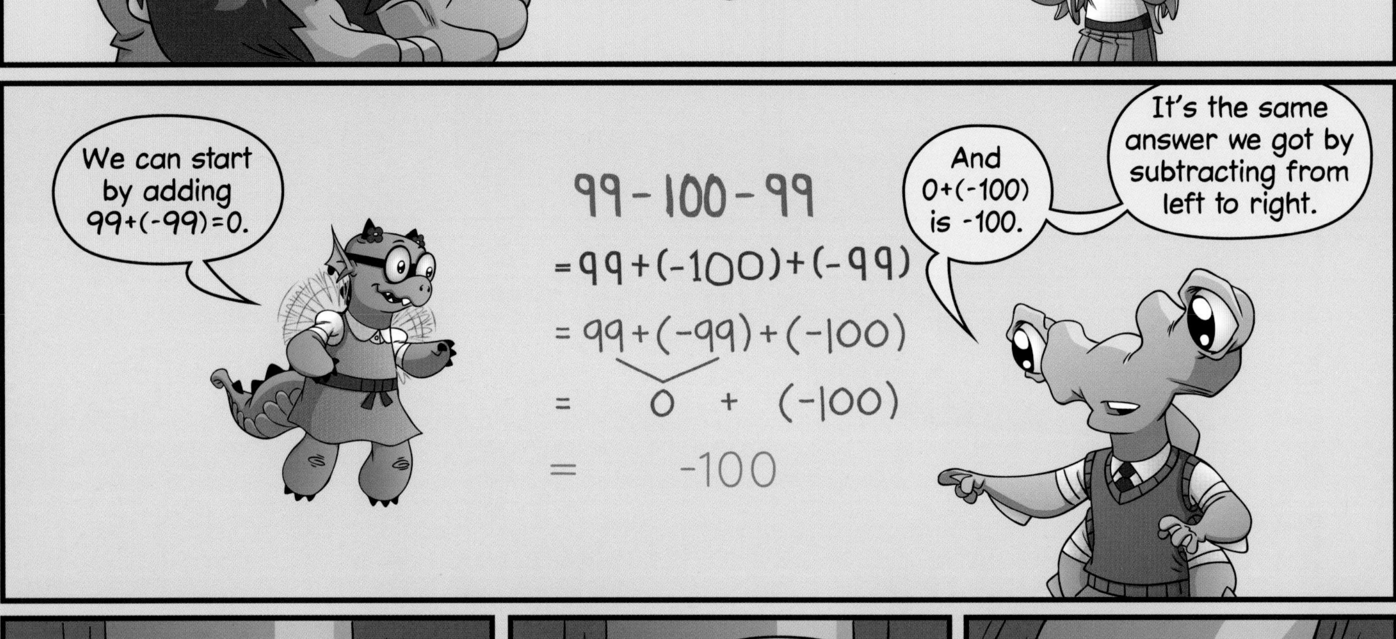
We can start by adding 99+(-99)=0.
99 - 100 - 99
=99+(-100)+(-99)
= 99+(-99)+(-100)
= 0 + (-100)
= -100
And 0+(-100) is -100.
It's the same answer we got by subtracting from left to right.

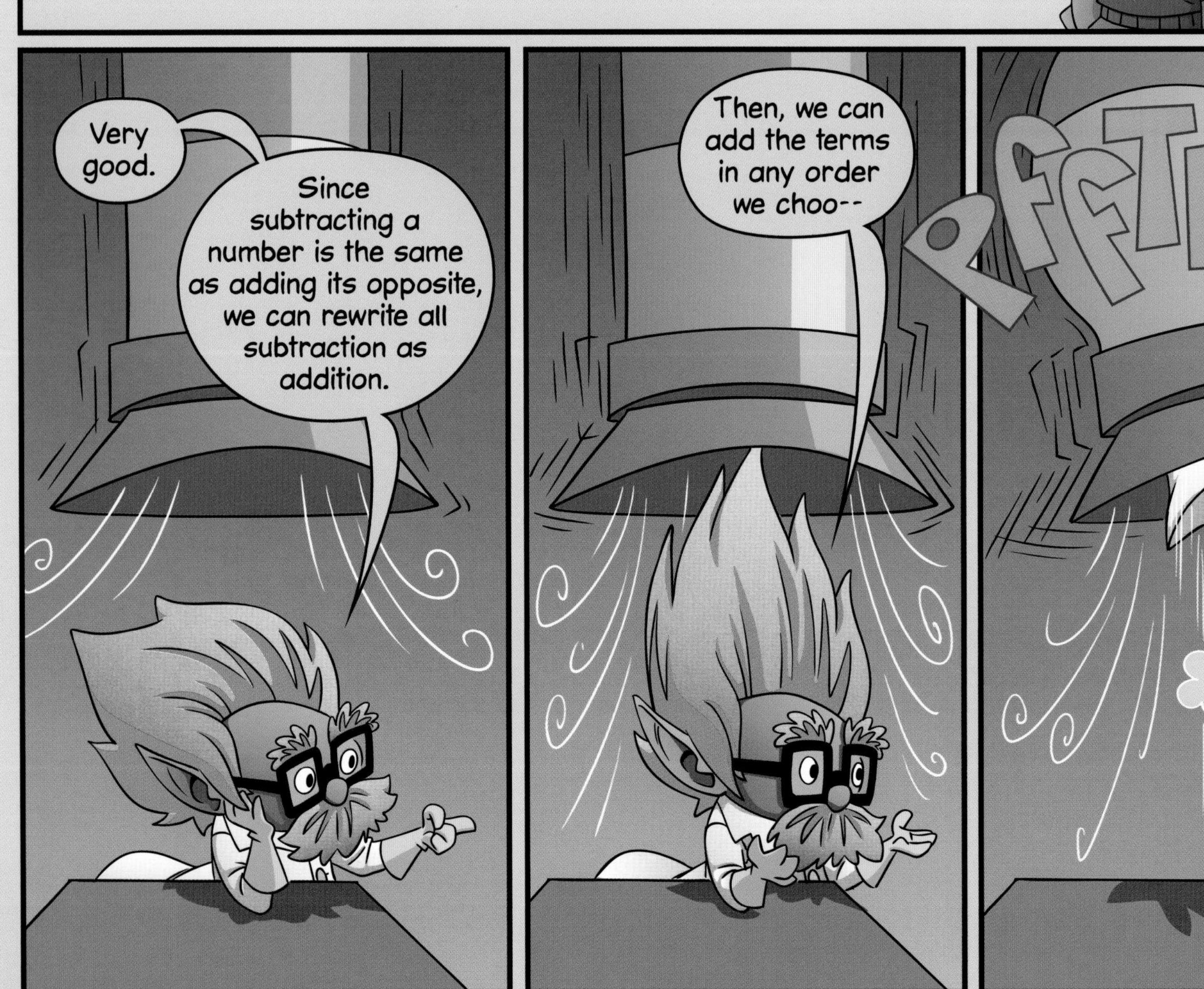
Very good.
Since subtracting a number is the same as adding its opposite, we can rewrite all subtraction as addition.
Then, we can add the terms in any order we choo--
PFFTHUMP

Bwah Hah Hah! Professor Grok is gone! I've abducted your educator! It's time for something diabolically difficult!
A summing of several integers is executed easily.
But simplifying a more substantial expression can be excruciating.
To save your schoolteacher, you must complete a clever computation.
Add each even integer from 2 to 1,000.
From that sum, subtract each odd integer from 1 to 999.
The result is the room number of your professor's imprisonment.
Compute quickly, or your educator will endure disquieting discomfort *indefinitely!*
PFFTHUMP
2+4+6+8+ ··· +996+998+1,000 − 1 − 3 − 5 − ··· − 995 − 997 − 999
That is a *lot* of numbers to add and subtract.
There must be a good way to do this.
Can you compute the value of Clod's expression?

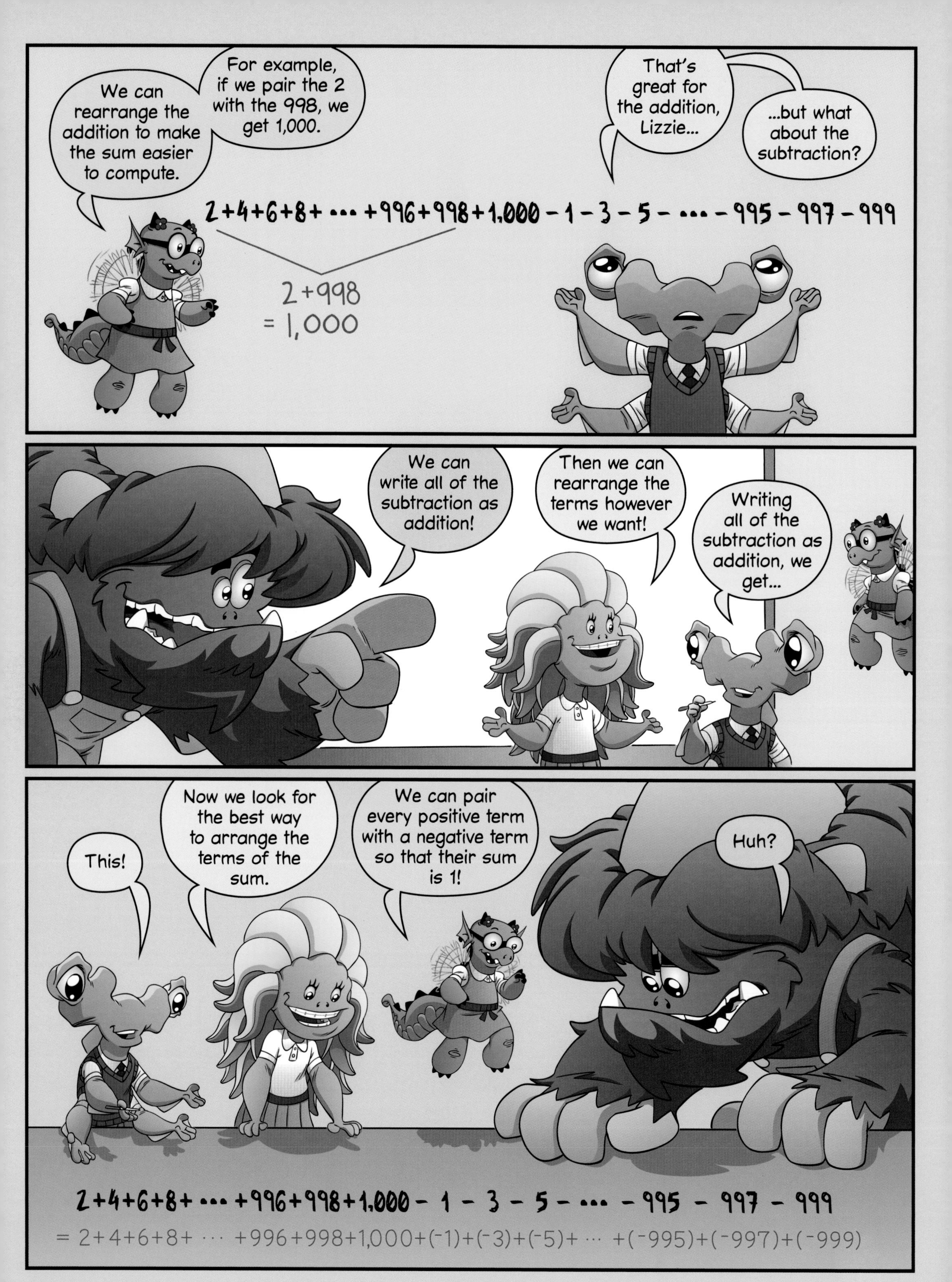
We can rearrange the addition to make the sum easier to compute.
For example, if we pair the 2 with the 998, we get 1,000.
That's great for the addition, Lizzie...
...but what about the subtraction?
2+4+6+8+ ··· +996+998+1,000 − 1 − 3 − 5 − ··· − 995 − 997 − 999
2+998 = 1,000
We can write all of the subtraction as addition!
Then we can rearrange the terms however we want!
Writing all of the subtraction as addition, we get...
This!
Now we look for the best way to arrange the terms of the sum.
We can pair every positive term with a negative term so that their sum is 1!
Huh?
2+4+6+8+ ··· +996+998+1,000 − 1 − 3 − 5 − ··· − 995 − 997 − 999
= 2+4+6+8+ ··· +996+998+1,000+(⁻1)+(⁻3)+(⁻5)+ ··· +(⁻995)+(⁻997)+(⁻999)

$$2+4+6+8+\cdots+996+998+1{,}000-1-3-5-\cdots-995-997-999$$
$$=2+4+6+8+\cdots+996+998+1{,}000+(^-1)+(^-3)+(^-5)+\cdots+(^-995)+(^-997)+(^-999)$$
$$=\underbrace{2+(-1)}+\underbrace{4+(-3)}+\underbrace{6+(-5)}+\cdots+\underbrace{996+(-995)}+\underbrace{998+(-997)}+\underbrace{1{,}000+(-999)}$$

$$2+4+6+8+\cdots+996+998+1{,}000-1-3-5-\cdots-995-997-999$$
$$=2+4+6+8+\cdots+996+998+1{,}000+(^-1)+(^-3)+(^-5)+\cdots+(^-995)+(^-997)+(^-999)$$
$$=\underbrace{2+(-1)}+\underbrace{4+(-3)}+\underbrace{6+(-5)}+\cdots+\underbrace{996+(-995)}+\underbrace{998+(-997)}+\underbrace{1{,}000+(-999)}$$
$$=1+1+1+\cdots+1+1+1$$

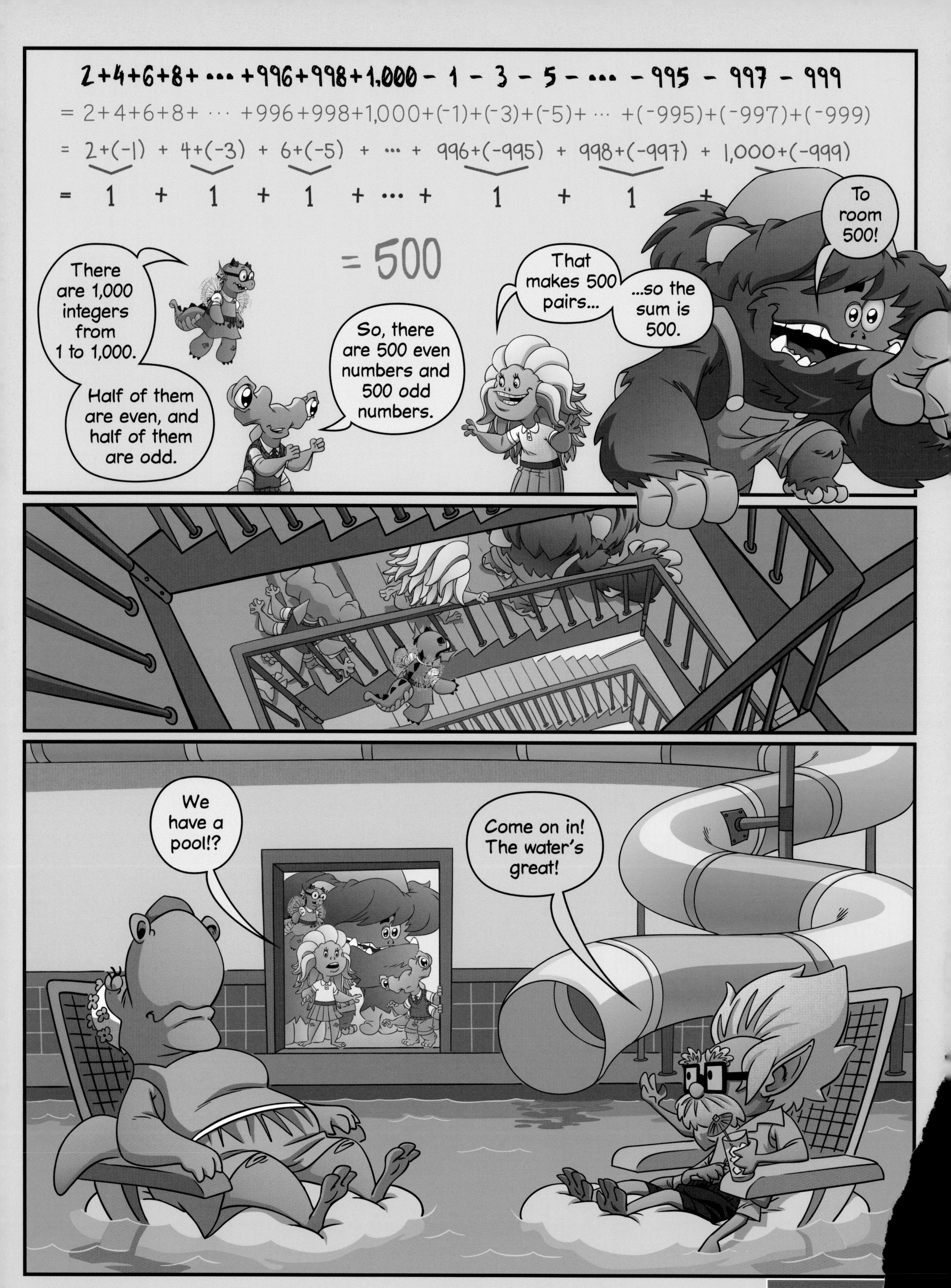
2+4+6+8+ ··· +996+998+1,000 − 1 − 3 − 5 − ··· − 995 − 997 − 999
= 2+4+6+8+ ··· +996+998+1,000+(−1)+(−3)+(−5)+ ··· +(−995)+(−997)+(−999)
= 2+(−1) + 4+(−3) + 6+(−5) + ··· + 996+(−995) + 998+(−997) + 1,000+(−999)
= 1 + 1 + 1 + ··· + 1 + 1 +
= 500
There are 1,000 integers from 1 to 1,000.
Half of them are even, and half of them are odd.
So, there are 500 even numbers and 500 odd numbers.
That makes 500 pairs...
...so the sum is 500.
To room 500!
We have a pool!?
Come on in! The water's great!

Writing Expressions without Negatives

Lizzie

Some expressions are a lot clearer if you write them without using negatives. Here are some ways to make your math a lot easier to read:

Subtracting a negative is the same as adding a positive.

$$a-(-b)=a+b$$

Examples:

$5-(-9)=5+9$ $7-(-2)=7+2$ $6-(-3)=6+3$

Adding a negative is the same as subtracting a positive.

$$a+(-b)=a-b$$

Examples:

$7+(-4)=7-4$ $10+(-3)=10-3$ $5+(-1)=5-1$

A negative plus a positive can be written as a difference between two positives.

☆ $$-a+b=b+(-a)=b-a$$

Examples:

$-2+6=6-2$ $-3+4=4-3$ $-11+5=5-11$

---

difference of two negatives can be written as a difference of two positives.

☆ $$-a-(-b)=-a+b=b-a$$

amples:

$-(-7)=7-2$ $-9-(-10)=10-9$ $-6-(-4)=4-6$

Index

Symbols

100 chart, 22–29

A

absolute value, 85–86
adding integers, 88–92
addition
- integers, 88–92
- mixed numbers, 64–66

associative property of addition, 101

C

Calamitous Clod. See Clod, Calamitous
Clod, Calamitous, 105
common factor, 53
commutative property of addition, 101
comparing
- fractions, 53–55
- integers, 83–86

composite, 20–29
converting
- fractions, 54–55
- fraction to mixed number, 60–61
- mixed number to fraction, 61–62

D

denominator, 50
divisibility, 32–39
- by 3, 38–39
- by 9, 32–39
- rules for 3 and 9, 39

divisible, 14

E

Eratosthenes, 30
exponents, 43

F

factoring. See prime factorization
factors, 12–47
- common, 53
- definition of, 14
- listing, 14–21
- pairs, 18
- prime, 40–47

factor tree, 41–45
Fiona. See Math Team
Fraction Jackson, 67
fractions, 48–77
- adding, 56–57, 73–77, 75–76
- comparing, 53–55
- converting, 54–55
- improper, 51
- on the number line, 50–52, 56–57
- simplest form, 53
- simplifying, 52–53
- subtracting, 68
- whole-number, 51

G

Grok. See Lab
Gym
- The Sieve, 22–30

H

homework, 67, 73–77

I

improper fractions, 51
index, 110–111
integers, 78–109
- adding, 88–92
- comparing, 83–86
- nonnegative, 93
- on the number line, 82–83, 88, 93
- ordering, 83–86
- positive, 93
- subtracting, 94–99

K

Kraken. See Woodshop

L

Lab
- Rearranging, 100–108

M

Math Team
- Adding Fractions, 56–57
- Adding Integers, 87–92
- Adding Mixed Numbers, 64–66

mixed numbers, 58–62
    adding, 64–66, 73–74
    subtracting, 69–72, 74–75
Ms. Q.
    Factors, 14–21
    Less Than Zero, 82–86
    Review, 50–55
multiples
    adding, 33
    on a 100 chart, 24–29

N

negative, 80–85
    integers, 93
    numbers, 80–85
    temperatures, 80–81
Notes
    Grogg's, 67, 93
    Lizzie's, 109
numerator, 50

O

opposites, 86
ordering integers, 83–86

P

positive, 20, 84
    integers, 93
prime. See prime number
prime factorization, 40–47
    exponents, 43
    factor trees, 41–45
prime number, 19
    1-digit, 19–21
    checking, 43–47
    even, 24
    listing, 22–30
    under 100, 29

ent and remainder, 59–61

uzz, 31
7/56, 63
g, 70–71

remainder, 34–37
    and mixed numbers, 59–61
R&G
    Divisibility, 32–39
    Mixed Numbers, 58–62
    Negatives, 80–81

S

School Bus
    Mental Math, 73–77
Sieve of Eratosthenes, 30
simplest form, 53
subtraction
    adding the opposite, 97
    fractions, 68–72
    integers, 94–99
    mixed numbers, 69–72, 74–75
    regrouping, 70–71
    writing as addition, 97–99, 104, 106

T

temperature, 80–81
tree. See factor tree

W

Woodshop
    Fraction Subtraction, 68–72
    Prime Factorization, 40–47
    Subtracting Integers, 94–99

For additional books,
printables, and more, visit
BeastAcademy.com